Advances in Microwave Processing for Engineering Materials

This text discusses recent research techniques in the field of microwave processing of engineering materials by utilizing microwave radiation in the form of microwave hybrid heating. It is useful for industrial and household applications including joining materials, casting bulk metal alloy material, drilling borosilicate glass materials, and developing cladding of different materials for friction, wear, and corrosion.

The book:

- Discusses the development of high-temperature resistant materials using microwave processing
- Covers the latest research development in microwave processing in the field of health care, for example, biomedical implants
- Highlights concepts of microwave heating in joining, cladding, and casting of metallic materials
- Explains mechanisms of failure of materials and protection in a comprehensive manner
- Provides readers with the knowledge of microwave processing of materials in major thrust areas of engineering applications

This book extensively highlights the latest advances in the field of microwave processing for engineering materials. It will serve as an ideal reference text for graduate students and academic researchers in the fields of materials science, manufacturing engineering, industrial engineering, mechanical engineering, and production engineering.

Advances in Microwave Processing for Engineering Materials

Edited by Amit Bansal and Hitesh Vasudev

CRC Press
Taylor & Francis Group
Boca Raton London New York

CRC Press is an imprint of the
Taylor & Francis Group, an **informa** business

First edition published 2023
by CRC Press
6000 Broken Sound Parkway NW, Suite 300, Boca Raton, FL 33487–2742

and by CRC Press
4 Park Square, Milton Park, Abingdon, Oxon, OX14 4RN

CRC Press is an imprint of Taylor & Francis Group, LLC

ISBN: 978-1-032-16478-6 (hbk)
ISBN: 978-1-032-16481-6 (pbk)
ISBN: 978-1-003-24874-3 (ebk)

DOI: 10.1201/9781003248743

Typeset in Sabon
by Apex CoVantage, LLC

Contents

Preface

In the competitive world, there is always a need for efficient, time-saving, and environmentally friendly manufacturing processes, which results in enhanced productivity with a low cost. In this direction, the use of microwave heating for processing of materials is gaining popularity because of the inherent features of microwave radiation, such as significant time savings associated with uniform heating and, more important, environmentally friendly processes. Microwave energy has emerged as a novel technique for processing engineering materials such as metal, ceramics, and composite materials in the form of joining, casting, melting, sintering, surface modifications in the form of cladding, and drilling of metal and ceramic-based materials with economy and ease. First and foremost, the current book covers the fundamental science and engineering aspects of microwave technology, the historical developments, microwave heating fundamentals, and its comparison with other conventional based heating processes, characteristics of microwave heating, different methods of generating the microwave radiation, general applications, advantages, and challenges.

Furthermore, technological progress has been addressed that includes the joining of materials, casting of bulk metal alloy material, drilling of borosilicate glass materials, development of cladding of different materials for friction, wear, and corrosion applications and the fabrication of composite materials. Recent advancements have been reported in high-temperature corrosion, erosion-resistant cladding, and functional graded cladding for many engineering applications.

The present book provides a state of the art in microwave energy by documenting the recent research trends that include the development of claddings of different materials for corrosion (hot corrosion and oxidation in boiler steels), friction and wear (hot forming dies), and biomedical (implants), among others. A newly introduced functional graded cladding of microwave processing of engineering components parts is discussed.

This book covers the studies and research works of reputed scientists and engineers who have developed the microwave energy for various engineering applications. Therefore, the book aims to compile the recent research trends occurring in the field of microwave processing of engineering materials by utilizing microwave radiation in the form of microwave hybrid heating technique for various industrial and household applications. Hence, the book will serve as a valuable resource for the fundamentals and the latest advancements in the field of microwave energy and as a consolidated reference for the professionals and aspirants of the researchers working in this area. Due care has been taken to explain and present the chapters with the help of self-explanatory schematic diagrams, data-enriched tables, structure–property correlations, and so on. We express sincere gratitude toward the contributing authors of different chapters of this book and the authors of different research articles, books, internet sites, and scientific reports referred to in this book.

Dr. Amit Bansal
Dr. Hitesh Vasudev

Editors' Biographies

Dr. Amit Bansal
I. K. Gujral Punjab Technical Univeristy, Kapurthala, Punjab, India-144603
https://orcid.org/0000-0002-0133-2897
https://ptu.ac.in/faculty/?fid=50

Dr. Amit Bansal is currently working as an assistant professor in the Department of Mechanical Engineering at I. K. Gujral Punjab Technical University, Kapurthala, India, since 2017. Earlier, he worked as an assistant professor at the Lovely Professional University from 2015 to 2017. He has obtained his M.Tech. (Production and Industrial Engineering) and PhD degrees from the Indian Institute of Technology (IIT) Roorkee, India. He has worked as a senior research fellow (SRF) during his PhD program at IIT Roorkee in a project sponsored by the Board of Research in Nuclear Sciences, Mumbai. The PhD topic was "Joining of Advanced Material Using Microwave Hybrid Heating Technique" at IIT Roorkee. His area of research is microwave processing of materials (joining, cladding, casting, and sintering) and surface engineering performed using thermal spray technology, plasma spray, and flame spray, among other topics. He has contributed extensively to the microwave processing of materials and thermal spray coatings in repute journals, which include publications with Elsevier, Taylor and Francis, Springer, IOP Publishing, and Intech Open. He has authored more than 50 research publications in various international journals, books, and conferences of repute. He has guided many master's and PhD scholars in thermal spraying, advanced manufacturing, and microwave processing of materials. He has teaching experience of more than eight years.

Dr. Hitesh Vasudev
Lovely Professional University, Phagwara, Punjab, India-144411
https://orcid.org/0000-0002-1668-8765
www.researchgate.net/profile/Hitesh_Vasudev2

Dr. Hitesh Vasudev is currently working as an associate professor at Lovely Professional University, Phagwara-India. He has received PhD in mechanical engineering from Guru Nanak Dev Engineering College, Ludhiana-India in 2018 (affiliated to Inder Kumar Gujral Punjab Technical University (IKGPTU)-Jallandhar). His area of research is thermal spray coatings, especially for the development of new materials used for high-temperature erosion and oxidation resistance and microwave processing of materials. He has contributed extensively in thermal spray coatings in repute journals, which include *Surface Coatings and Technology*; *Materials Today Communications*; *Engineering Failure Analysis*; the *Journal of Cleaner Production, Surface Topography: Metrology and Properties*; the *Journal of Failure Prevention and Control*, the *International Journal of Surface Engineering and Interdisciplinary Materials Science* under the flagship of various publication groups such as Elsevier, Taylor & Francis, Springer nature, IGI Global and Intech Open. Moreover, he is a dedicated reviewer of reputed journals such as *Surface Coatings and Technology, Ceramics International*, the *Journal of Material Engineering Performance, Engineering Failure Analysis, Surface Topography: Metrology and Properties Material Research Express, Engineering Research Express*, and IGI Global journals, among others. He has authored more than 30 International publications in various international journals and conferences. He has published 15 book chapters in various books related to surface engineering and manufacturing processes. He has also published a unique patent in the field of thermal spraying. He has teaching experience of more than eight years. He received a "Research Excellence Award" in 2019, 2020, and 2021 at Lovely Professional University, Phagwara, India. He has organized a National Conference and has been a part of many international conferences.

List of Contributors

Dinesh Agrawal
The Pennsylvania State University
State College, Pennsylvania, USA

Ramkishor Anant
Maulana Azad National Institute of Technology
Bhopal, Madhya Pradesh, India

Amit Bansal
I. K. Gujral Punjab Technical University
Jalandhar, India

T. Lachana Dora
Birla Institute of Technology and Science
Pilani, India

Pranjal Gupta
Indian Institute of Technology
Roorkee, Uttarakhand, India

Rahul Gupta
Indian Institute of Technology
Roorkee, Uttarakhand, India

Shivani Gupta
Indian Institute of Technology
Roorkee, Uttrakhand, India

Ajit M. Hebbale
N.M.A.M. Institute of Technology
Udupi Dist, Karnataka, India

Eli Jerby
Faculty of Engineering
Tel Aviv University, Israel

Sarbjeet Kaushal
Gulzar Group of Institutions
Khanna, Ludhiana, India

Dinesh Kumar
Rayat Bahra University
Mohali, Punjab, India

Gaurav Kumar
National Institute of Technology
Uttarakhand, India

Rajeev Kumar
Indian Institute of Technology
Mandi, Himachal Pradesh, India

Mahantayya Mathapati
KLE College of Engineering & Technology
Karnataka, India

Radha Raman Mishra
Birla Institute of Technology and Science
Pilani, India

Himanshu Pathak
Indian Institute of Technology Mandi
Mandi, Himachal Pradesh, India

Guru Prakash
Maulana Azad National Institute of Technology
Bhopal, Madhya Pradesh, India

C. Durga Prasad
Department of Mechanical Engineering
RV Institute of Technology and Management
Bengaluru 560076 Karnataka, India

Gaurav Prashar
School of Mechanical Engineering
Lovely Professional University, Punjab-India

M.R. Ramesh
National Institute of Technology Karnataka
Mangalore, India

Manjeet Rani
Indian Institute of Technology Mandi
Mandi, Himachal Pradesh, India

Apurbba Kumar Sharma
Indian Institute of Technology
Roorkee, Uttarakhand, India

Anurag Singh
Indian Institute of Technology Roorkee
Uttarakhand, India

Jashanpreet Singh
Punjab State Aeronautical Engineering College
Punjab, India

M.S. Srinath
Malnad College of Engineering
Karnataka, India

Gudala Suresh
National Institute of Technology Karnataka
Surathkal, Mangalore, India

Lalit Thakur
National Institute of Technology
Kurukshetra-India

Nishant Verma
Indian Institute of Technology
Mandi, Himachal Pradesh, India

Hitesh Vasudev
School of Mechanical Engineering
Lovely Professional University, Punjab-India

Rahul Yadav
Punjab State Aeronautical Engineering College
Patiala, Punjab, India

Sunny Zafar
Indian Institute of Technology Mandi
Mandi, Himachal Pradesh, India

Chapter 1

Microwave Processing of Materials

Fundamentals and Applications

T. Lachana Dora and Radha Raman Mishra

Contents

1.1 Introduction

In recent years, manufacturing strategies for producing the components in industries, such as aerospace, automobiles, and nuclear reactors, have

DOI: 10.1201/9781003248743-1

changed significantly due to the use of advanced materials and stringent manufacturing guidelines across the globe for energy-saving and eco-friendly manufacturing. Interaction of heat with the advanced materials during the manufacturing stage significantly influences the microstructural characteristics of the objects and properties of the products. In the last decade, microwave energy has demonstrated more uniform heating characteristics during the processing of various materials. The properties of microwave energy processed products depend on the effective heat generation/heat transfer during microwaves–material interactions. In the following sections, various aspects of electromagnetic spectrum and microwaves have been discussed.

1.2 Microwaves

Electromagnetic (EM) radiation travels at the speed of light and interacts with material in waveform. It consists of electric and magnetic fields that oscillate perpendicular to each other in the direction of propagation. The electromagnetic theory was proposed by James Clerk Maxwell to establish the fact that electromagnetic radiation is transmitted in waves or particles of different wavelengths and frequencies. Different wavelengths form the EM spectrum, which consists of seven domains with decreasing wavelengths and increasing energy and frequency. EM waves include radio waves ranging from the 'long-wave' band through VHF, UHF, and beyond; microwaves; visible, infrared, and ultraviolet light; X-rays and gamma rays; and so on (Figure 1.1). Microwaves exist between radio and infrared light in the EM spectrum.

Microwaves are EM waves that have wavelengths starting from one mm to one thousand mm and frequencies ranging from 300 MHz to 300 GHz. Cellular phones, radar, and TV satellite communications work within this region of the spectrum [1].

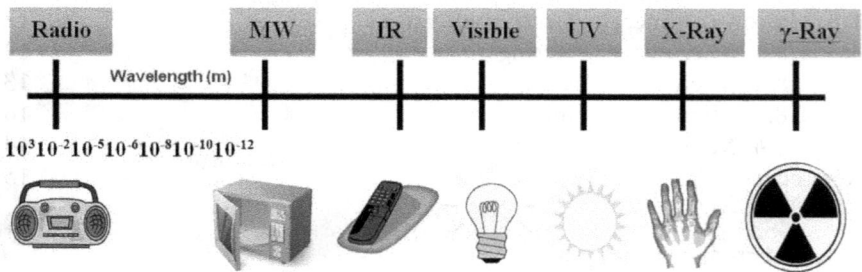

| Radio | | MW | IR | Visible | UV | X-Ray | γ-Ray |

Wavelength (m)

$10^3 10^{-2} 10^{-5} 10^{-6} 10^{-8} 10^{-10} 10^{-12}$

Figure 1.1 EM spectrum.

1.2.1 Historical Background

Michael Faraday proposed in 1832 that electromagnetic phenomena, like sound and light, are associated with wave motion. Maxwell devised mathematical models in 1846 to comprehend and explain electromagnetic phenomena, such as the interaction of the electric and magnetic fields with matter. Heinrich Hertz carried out several experiments to demonstrate that Maxwell's equations accurately described the electromagnetic wave propagation across the atmosphere and space. In 1888, he demonstrated that radio waves could be rebounded off things in the same way that sound waves could. In 1903, Hulzmeyer patented a system for detecting obstructions and navigating ships using reflected radio signals. Clavier of ITT builds an 18-cm-wavelength radio connection from Calais, France, to Dover, England, in 1931. The micro-ray was another technique that utilized a signal so-called Barkhausen tube. This term was crucial in describing this part of the spectrum as "microwaves" [2]. In the United Kingdom (1939), Boot and Randall invented the cavity magnetron. The tube produced 1-kW pulsed power and had a 10-cm wavelength. During World War II, this device was crucial in the development of microwave radar in the United States. Percy Spencer of Raytheon disclosed fundamental microwave oven and cooking inventions in 1949, including the utilization of a cooking cavity and food preparation methods [3]. Since then, microwave energy started being using for different processes, which can be seen in Figure 1.2.

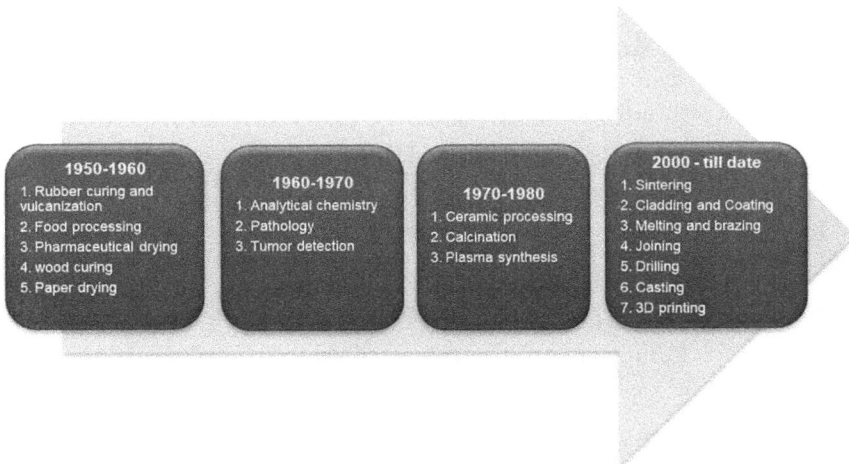

1950-1960
1. Rubber curing and vulcanization
2. Food processing
3. Pharmaceutical drying
4. wood curing
5. Paper drying

1960-1970
1. Analytical chemistry
2. Pathology
3. Tumor detection

1970-1980
1. Ceramic processing
2. Calcination
3. Plasma synthesis

2000 - till date
1. Sintering
2. Cladding and Coating
3. Melting and brazing
4. Joining
5. Drilling
6. Casting
7. 3D printing

Figure 1.2 Chronological development in microwave processing.

1.3 Microwave Heating Fundamentals

As discussed earlier, microwaves are a portion of the electromagnetic spectrum with wavelengths ranging from 1 mm to 1 m and frequencies ranging from 300 MHz to 300 GHz. The frequencies 0.915 GHz and 2.45 GHz are commonly used for microwave material processing. These frequencies are reserved by the Federal Communications Commission (FCC) for heating purposes in industry, scientific experiments, and medical (ISM) work [1]. The interactions of microwaves with a material are dependent on dielectric and magnetic responses of the material as the electric and magnetic fields influence the material during irradiation.

1.3.1 Microwave Energy–Based Heating Devices

The source, transmission lines, and the applicator are the three main components of a microwave furnace. The microwave source generates electromagnetic radiation, which is then supplied to the applicator through transmission lines.

1.3.2 Microwave Source

Most of the microwave heating industry requires a large amount of heat which needs a power rating of 10 KW to 100 KW. To generate this amount of power, the generator efficiency must be high with less wastage of energy. Furthermore, the generated power must be of stable frequency and devoid of harmonics and other spurious frequencies in order to comply with international frequency allocation standards [4]. The most popular microwave sources are mentioned in Figure 1.3.

In recent times, although other microwave power sources are developed, still, most microwave heating systems use a magnetron as their power source due to the fact that magnetrons are mass-produced in large quantities and are the most cost-effective sources.

1.3.3 Magnetron

Magnetrons, first created in 1921 and with much improved versions of it being developed until 1940, produce continuous or pulsed microwaves with

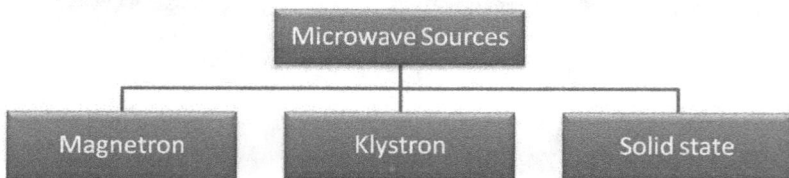

Figure 1.3 Types of microwave sources.

energy up to megawatts and frequencies ranging from 1 GHz to 40 GHz. The efficiency of magnetron-based devices is about 80%, with a life span beyond 5000 hours. A magnetron is a cross-field device in which magnetic and electric fields are applied perpendicular to each other. The generation of microwaves involves an interaction of EM waves with electrons traveling in the crossed fields. In a magnetron, a cathode is surrounded by an anode block, and the space between the anode and the cathode called the interaction space, whereas the spaces available with magnetron are known as resonant cavities that determine the output frequency of the microwave. Electric and magnetic fields are generated by resonant cavities (known as tuned circuits) and a magnet surrounding the magnetron, respectively. The magnetron features an antenna that extends into resonant cavities and the antenna is coupled with the anode. Mutually perpendicular electric and magnetic fields are generated when electricity passes through the electrode and a magnetic field is applied parallel to the axis. The electrons at the cathode accelerate radially initially; afterward, they follow the cycloidal paths owing to the presence of the magnetic field. The strong magnetic field forms a spinning space charge by impeding the electrons to reach the anode. The resonant cavities of the anode interact with electrons by either accelerating or deaccelerating them. Consequently, microwave frequencies cause electron bunches to flow around the cathode, which results in self-sustaining oscillations of the resonant cavities. The microwaves are transmitted from the magnetron to the waveguide and the coupling loop extracts some of the microwave power.

1.3.4 Transmission Line

The microwave energy is transferred to the applicator cavity from the source via the transmission lines. A waveguide, a hollow metal conduit with rectangular/cylindrical cross sections, is popularly used for electromagnetic waves travel. Rectangular waveguides are generally utilized in industrial microwave heating; however, different waveguides with specific shapes can be created for specific applications [5]. In waveguides, microwaves can propagate can in two forms: transverse electric (TE: electric force in direction of propagation is zero) and transverse magnetic (TM: magnetic intensity is zero in the direction of propagation). The mathematical solution of an electromagnetic wave in a rectangular waveguide can be analyzed into a linear combination of TE and TM modes. Apart from a waveguide, there are some other components that assist in equipment protection, detection, and microwave coupling with the material in the applicator cavity.

1.3.5 Circulators

Microwave energy reflects back to the microwave source while irradiating bad microwave absorbers; consequently, magnetrons get often damaged. The circulator protects the microwave instrumentation by behaving as a sort of a diode in an electrical circuit. The circulator could be a three-port waveguide part with nonreciprocating options in terms of forward and reflected wave treatment. Microwaves will solely travel in a way via the circulator. The three ports of the circulators are connected with the microwave source, the applicator, and a faux load. The arrangement deflects the reflected microwave power, and the mirrored power gets absorbed by a dummy load that is sometimes water [1, 5].

1.3.6 Applicator

An applicator is a device that irradiates microwave radiation into a targeted material volume at an appropriate intensity, which changes material attributes permanently or temporarily. Such changes may include an increase in temperature, elimination of moisture, amplification of a chemical process, ablation of biological tissue, breakdown of gases to form plasmas, heating of metals and ceramics, and so on. Applicator cavities can be of different shapes and sizes depending on the material processing requirements; however, all applicators essentially consist of a void surrounded by high-conductivity metal walls. The classification of microwave applicators is presented in the following sections.

1.3.6.1 Single Mode Applicator

The applicators which create just one mode are known as single-mode applicator [1, 6]. It enables researchers to separate the electric and magnetic fields with correct tuning. It provides flexibility to study the effects of E and H fields on various classes of materials individually. In a single-mode cavity, the materials being processed or sensed should be placed in areas with the highest field intensity node. An analytical or numerical approach can be used with necessary boundary conditions for a given applicator geometry. Thus, theoretical analysis of the microwave applicator/cavity using the Maxwell equations can be used to characterize the behavior of microwaves. Single-mode applicators are made up of a section of waveguide that works near the cutoff frequency (Figure 1.4).

The design of single-mode applicators supports one resonant mode that provides higher microwave field intensity precisely at a given location inside the applicator cavity. Transverse electric (TE or H) and transverse magnetic (TM or E) modes are frequently used in single-mode cavities having a rectangular or circular cross section [7]. However, single-mode applicators have various drawbacks, such as product-specific uses rather than general-purpose use; being extremely sensitive (i.e., tuned off-resonance) to changes in product attributes, geometry, and position; and are expensive. Some popular

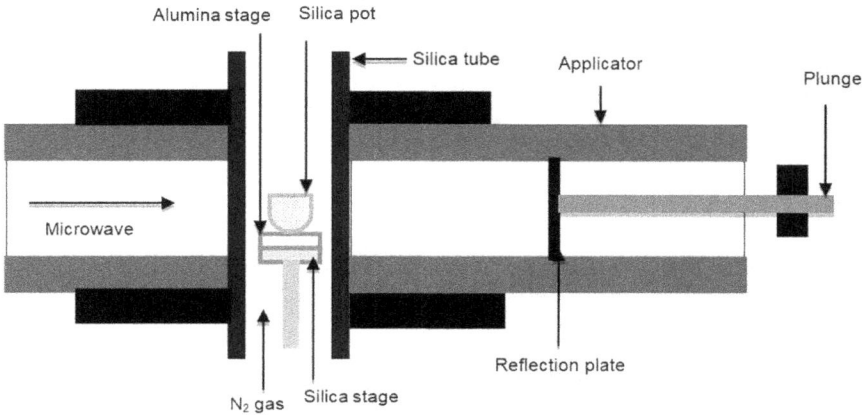

Figure 1.4 A schematic diagram of the single-mode applicator.

applications of single-mode applicators include materials processing with phase changes, mineral processing, sintering of ceramic materials, and so on.

1.3.6.2 Multimode Applicator

Multimode cavities are applicators that can support many high-order modes simultaneously. There are many active modes within a multimode cavity, and hence, it is hard to separate the electric and magnetic fields at a specific location inside the cavity. Multimode microwave applicators include a wide cavity that encloses a volume and is completely encircled by conducting wall. Many modes resonant in the area of the generator's operating frequency are required in a resonant multimode oven. The homogeneity of microwave field intensity can be enhanced by increasing the cavity size or operating at a higher frequency. At higher frequencies, a smaller size applicator cavity offers a more uniform distribution of microwave field intensity due to shorter wavelengths. Therefore, multimode ovens are operated at 2.45 GHz to achieve more uniformity. Additionally, the mode stirrer or turntable can further improve the uniformity inside the cavity. The mode stirrers, reflectors placed near the waveguide input that reflect microwaves waves off irregularly shaped objects, redistribute the electromagnetic field intensities inside the cavity. Therefore, multimode applicators are more popular as compared to single-mode applicators due the flexibilities they offer in material processing.

Mode quality factor (Q-factor): The quality factor of a simple resonant circuit indicates its damping and the frequency bandwidth of its response. The Q-factor can be represented as

$$Q = 2\pi \frac{total\,energy\,stored}{energy\,dissipated\,/\,cycle}$$

$$= 2\pi \frac{total\,energy\,stored \times resonant\,frequency}{power\,dissipated} \tag{1.1}$$

$$Q = \frac{f_0}{\delta f}, \tag{1.2}$$

where f_0 is the resonant frequency and δf is the total frequency spread between the half-power points on the frequency response curve [8].

1.4 Maxwell's Equations and Dielectric Response

Maxwell proposed a set of mathematical formulations to better understand the combined effects of the E and H field intensities. The complexities of the electromagnetic field have so far eluded easy explanation, and much more research is needed to fully comprehend the various phenomena. By adding proper boundary conditions to the equations, the microwave–materials interactions may be explored. Maxwell's equations are given as [9, 10]

$$\nabla \times E = j\mu\omega H \tag{1.3}$$

$$\nabla \times H = -j\omega\varepsilon_0\varepsilon^* E \tag{1.4}$$

$$\nabla \cdot (\varepsilon E) = 0 \tag{1.5}$$

$$\nabla \cdot H = 0, \tag{1.6}$$

where E, H, ε^*, ε, ε_0, μ, ω, and j are time-harmonic electric field, time-harmonic magnetic fields, the complex permittivity of the material, the permittivity of the material, the permittivity in air, the magnetic permeability, the angular frequency equal to $2\pi f$ (f = is frequency of microwaves), and the imaginary unit, respectively.

The material interaction with E and H fields is based on the dielectric, electric, and magnetic properties of the material. The influence of E field on the material characteristics can be analyzed by considering the complex permittivity (ε^*), which quantifies the material's tendency to polarize and the losses inside the material due to the presence of a sinusoidal electric field. The complex permittivity of a material is expressed as

$$\varepsilon^* = \varepsilon' - j\varepsilon'', \tag{1.7}$$

where ε' and ε'' are the absolute permittivity (indicates penetration of microwaves into the material) and the dielectric loss factor (represents the ability of the material to store energy), respectively. The materials capacity to convert absorbed energy into heat is often analyzed by the loss tangent (*tan δ*), which is expressed as [1, 11]

$$\tan \delta = \frac{\varepsilon''}{\varepsilon'}. \tag{1.8}$$

The angle (*δ*) measures the phase difference between the **E** field and the polarization of the material.

1.5 Penetration Depth and Power Absorbed

The heating of a material during irradiation happens due to the absorption of microwave energy inside the material; as a result, intensity of the field that is more likely to interact with the material diminishes as microwaves passes through a material [4, 5, 8]. The field intensity and accompanying power flux density drop exponentially as microwave penetrates inside a perfectly dielectric material. Power dissipation from the surface decreases exponentially because the power absorbed in an elemental volume of material is proportional to the power flux density passing through it. The penetration depth D_p, or skin depth has a significant impact on the power absorption and is governed by relative permittivity and the loss factor of the material. Mathematically, the depth inside the material (the surface of the material is taken as a reference) at which the magnitude of the field strength remains $1/e$ (= 0.368) of its surface value is known as the penetration depth of the field [9, 12, 13]

$$D_p = \frac{1}{\alpha}, \tag{1.9}$$

where α is called the attenuation factor and generally termed as

$$\alpha = \omega \frac{\sqrt{\mu_0 \mu' \varepsilon_0 \varepsilon'}}{\sqrt{2}} \sqrt{\left(\sqrt{1 + \left(\frac{\varepsilon''_{eff}}{\varepsilon'} \right)^2} - 1 \right)}. \tag{1.10}$$

For absorbing medium, where $\varepsilon''_{eff} \gg \varepsilon'$, α reduced to

$$\alpha = \sqrt{\frac{\omega^2 \mu' \mu_0 \varepsilon''_{eff} \varepsilon_0}{2}}. \tag{1.11}$$

For a transparent medium, where $\varepsilon''_{eff} \ll \varepsilon'$, α reduced to

$$\alpha = \frac{\omega}{2} \sqrt{\frac{\mu' \varepsilon_0 \mu_0}{\varepsilon'}} \varepsilon''_{eff}. \tag{1.12}$$

In the case of conducting material, the penetration depth is termed the skin depth and defined as

$$D_s = \sqrt{\frac{2}{\sigma \omega \mu}}, \tag{1.13}$$

where α is $1/D_s$ and σ is the electrical conductivity (S/m).

The conversion of microwave energy into heat depends on dielectric, electric, and magnetic responses of the target material. Theoretical analyses of microwave heating of a material can be described by the equations of the power density distribution [4, 9]. The average power dissipation due to electric losses is

$$\left(P_{avg}\right)_{electric} = 2\pi f \epsilon_0 \varepsilon''_{eff} E^2_{rms}. \tag{1.14}$$

Similarly, the average power dissipation due to magnetic losses is

$$\left(P_{avg}\right)_{magnetic} = 2\pi f \mu_0 \mu''_{eff} H^2_{rms}. \tag{1.15}$$

Combined, the total power dissipation is termed as

$$\begin{aligned}\left(P_{avg}\right)_{total} &= \left(P_{avg}\right)_{electric} + \left(P_{avg}\right)_{magnetic} \\ &= 2\pi f \epsilon_0 \varepsilon''_{eff} E^2_{rms} + 2\pi f \mu_0 \mu''_{eff} H^2_{rms}\end{aligned} \tag{1.16}$$

where, E_{rms} and H_{rms} denote to the root mean square of the electric field and the magnetic field, respectively; ε''_{eff} denotes the effective relative dielectric loss factor; and μ''_{eff} denotes the effective relative magnetic loss factor.

1.6 Microwave–Material Interaction

Microwaves interactions with a targeted material depends upon the response of material to the **E** and **H** field components. Depending up on the material characteristics, the microwaves can be reflected, transmitted, or absorbed by the material/its constituents. The different possible cases of microwave–material interaction with relevant materials are presented in Table 1.1. The case of the mixed absorber highlights the selective heating characteristics of the microwaves.

1.7 Characteristics of Microwave Heating

Microwave heating of materials differs greatly from traditional heating. Energy gets dissipated volumetrically at the molecular/atomic level due to conversion of wave energy into heat; however, the transfer of heat from outer surfaces of the object to its core happens in conventional heating. The same fact is illustrated in Figure 1.5.

Heat is transported into the material from the outside to the interior by conduction, convection, and radiation in typical conventional heating. Therefore, in conventional heating, the temperature of the core will be less

Table 1.1 Possible Microwave–Material Interactions

Interaction	Schematic	Relevant Materials
Transparent (low dielectric loss)		Teflon and quartz
Absorber (high dielectric loss)		Water and SiC
Opaque (no dielectric loss)		Bulk metals
Mixed absorber (variation in dielectric loss)		Polymer matrix composites, Ceramic matrix composites Metal matrix composites

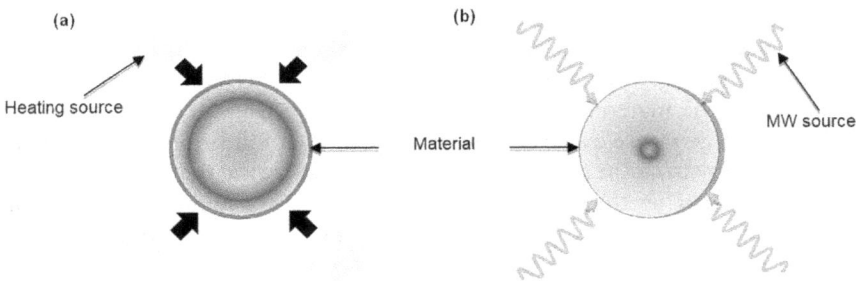

Figure 1.5 Temperature distribution pattern in (a) conventional heating (b) microwave heating.

compared to surface. In contrast, microwaves heat a material by converting the wave energy into heat throughout the entire volume (volumetric heating). The heat flow in microwave heating is directed from the core to the outer surface of the material [14]. The temperature of microwave irradiated materials is higher at the core as compared to the temperature at the surface. Volumetric heating offers various advantages including reduced processing time, reduced energy usage, more uniform heating, and enhanced diffusion rate. The following approaches have been reported for processing different material systems using microwave irradiations [9, 15].

1.7.1 Direct Heating

In case of direct heating, materials are directly placed in the microwave cavity and are exposed to microwaves. Microwave energy is absorbed by the substances with subsequent heating. Microwave direct heating is the most commonly used on materials that are readily coupled to microwaves, such as lossy ceramics, micro-nanometallic powders, food products, and so on. The intrinsic temperature differential induces overheating of the material. It may result in the formation of hot spots and subsequent formation of thermal runway during direct microwave heating.

1.7.2 Hybrid Heating

A large portion of the available materials don't promptly couple to microwaves at room temperature. The hybrid heating approach was proposed for microwave heating of such materials. Hybrid microwave heating entails two-step heating: (1) *conventional heating*—to enhance microwave coupling of the materials by raising the temperature of the target material beyond its critical temperature and (2) *self-microwave absorption*—beyond critical temperature material couples with microwaves and rapid microwave heating of material occurs. A first step can be accomplished by transferring heat to the material using an outside source (e.g., electric furnace) or microwave susceptors that couple with microwaves at room temperature gets heated quickly to act as a source of heat for the target material. Initially, heat is transferred from the susceptor to the material by conduction and convection; however, when the susceptor temperature exceeds a certain value, radiation begins. The material absorbs energy from the susceptor until it reaches a critical temperature. When the material reaches the critical temperature, it absorbs microwave radiation directly and generates fast internal heat. The heat is moved from the inside center to the outside surface of the material during this stage.

1.7.3 Selective Heating

Heating materials selectively is a unique property of microwaves. It offers flexibility in heating a specific material or a specific region of material while irradiating a material system or a single-phase material. The selective heating is driven by the fact that the difference in the dielectric or magnetic response of a multi-material system. Microwaves gets attracted by materials whose dielectric loss factor/magnetic loss factor are higher; consequently, these materials receive rapid heating. This characteristic of microwaves provides opportunities to explore novel methods or development of materials with new or distinctive microstructures, synthesize functionally hierarchical materials (FGMs) and to affix ceramic materials [16].

1.8 Microwave Energy–Based Applications

1.8.1 Microwave Sintering

Microwave energy–based sintering is a novel process that was developed in the mid- to late 1980s. In the early stage, the process was known as rapid densification of ceramic materials. Multimode industrial sintering furnaces or microwave ovens operating at 2.45 GHz can be used for rapid heating of green components (up to 100°C/min). Single-mode microwaves are useful for sintering poor microwave absorbers or sintering a material at higher heating rates. A schematic diagram of a microwave sintering setup is shown in Figure 1.6. Microwaves get concentrated on the green component; it allows better power density for rapid microwave sintering of the component [17].

In contrast to traditional sintering, the energy in the microwave sintering process is transmitted directly to the sample. As a result, rapid volumetric heating occurs, which leads to significant a decrease in sintering time and energy consumption up to three times of that required for conventional sintering. In earlier days, most of the studies in microwave sintering was carried out on ceramics, such as SiO_2, B_4C, Al_2O_3, TiO_2, ZrO_2, and ZnO, among others. However, after the evolution of the hybrid microwave heating approach, sintering of ceramic composites, metal matrix composites (such as AMMC, Mg-MMC, Ti-MMC, etc.) [18], titanium-based composite for orthopedic applications [19], and titanium-based shape memory alloy (Ti-23%Nb) [20] were reported with improved mechanical properties.

1.8.2 Microwave Casting

Microwave casting is a recently explored area of microwave applications. Microwave casting can be classified into two types: in situ (charge melting and casting are accomplished inside the microwave applicator cavity) and ex situ (only charge melting is carried inside the microwave applicator cavity) microwave casting. The schematic of the in situ microwave casting is shown in Figure 1.7. A pouring basin is used to melt the charge, and the melt passes through a sprue. The sprue guides the molten metal to fill inside the mold cavity and subsequent solidification. In contrast, in the ex situ microwave casting method, the charge is melted using microwave irradiations and solidification of the melt is done outside the cavity. The ex situ microwave casting

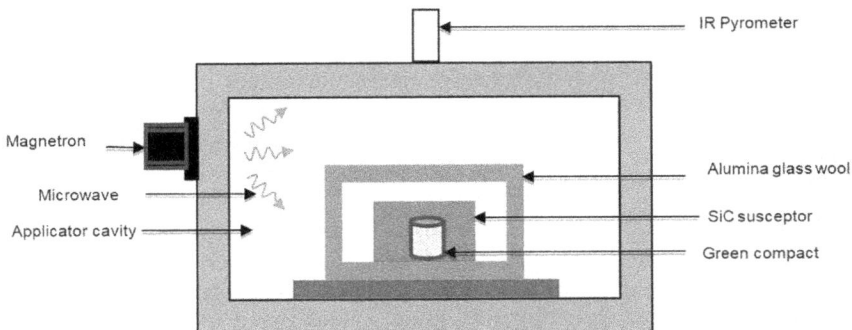

Figure 1.6 A schematic of the microwave sintering setup.

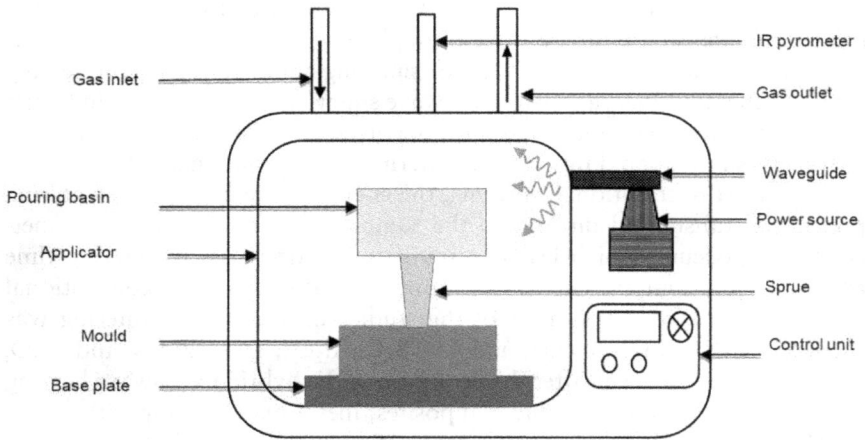

Figure 1.7 Schematic of microwave casting process.

resembles the conventional casting process and lacks benefits like control over the mold preheating temperature, molten metal flow, and grain growth as achieved in the in situ microwave casting [21]. The process was used for casting of Al alloy [22–25] and Ni-based metal matrix composite [26–27].

1.8.3 Microwave Joining

The joining of two similar and dissimilar bulk/pipe pieces can be accomplished using microwave radiation. A schematic diagram of the microwave joining setup is shown in Figure 1.8. A hybrid approach of heating the base metal using a susceptor was used for ceramic and metal joining [28]. A filler powder of base metal or different compatible metal is used for joining the pieces. A separator is used to avoid diffusion of contaminants from the susceptor layer into the welded area. A refractory brick is used as fixture to accommodate the pieces that are to be welded. In recent years, joining various materials such as SS-316 to MS plates [29], cast iron joints [30], MS pipes [31], and SS202 sheets [32] with improved properties has been reported.

1.8.4 Microwave Cladding

Microwave cladding is the partial melting/sintering of a preplaced clad thin layer on a substrate material using microwave irradiation. A schematic diagram of the cladding setup is shown in Figure 1.9. In the microwave cladding process, the polished metallic substrate is covered with clad powder. A separator material is used to avoid diffusion between the susceptor and the

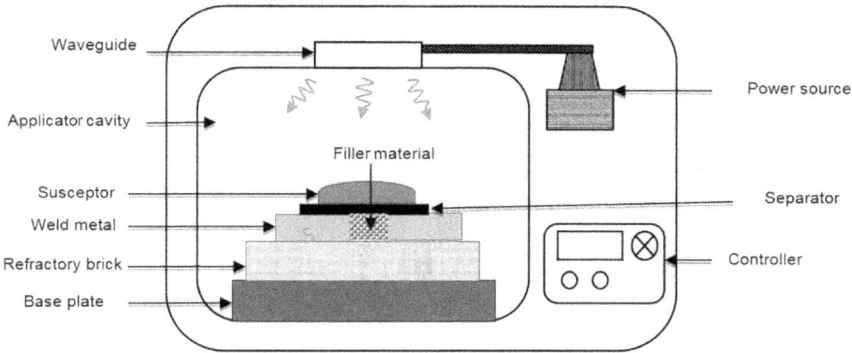

Figure 1.8 A schematic diagram of the microwave joining setup.

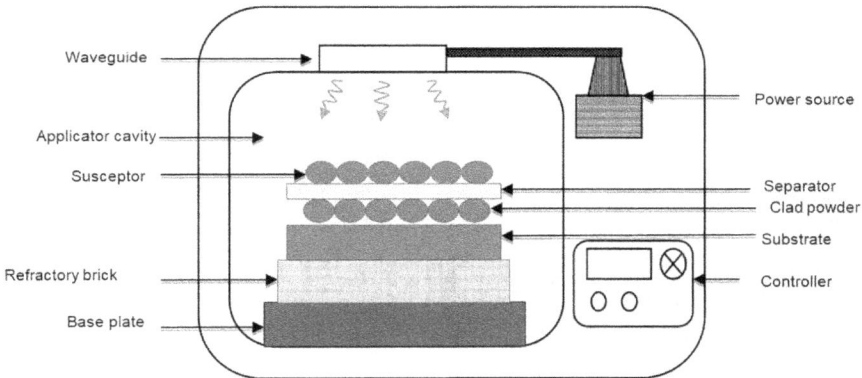

Figure 1.9 A schematic diagram of the microwave cladding setup.

clad layer. The separator absorbs microwave energy and hybrid heats the clad powder. It also allows microwaves to interact directly with clad powder.

In the recent years, microwave cladding of different material systems, such as SS-WC10Co2Ni powder [33], CA6NM hydro-turbine steel-Ni-based alloy powder [34], SS-420 substrate-nickel-based composite powder [35], and SS-420 substrate-cobalt-based clad [36], with improved properties was reported.

References

[1] E. T. Thostenson, and T. W. Chou, "Microwave processing: Fundamentals and applications," *Compos. Part A Appl. Sci. Manuf.*, vol. 30, no. 9, pp. 1055–1071, 1999, doi: 10.1016/S1359-835X(99)00020-2.

[2] H. Sobol, and K. Tomiyasu, "Milestones of microwaves," *IEEE Trans. Microw. Theory Tech.*, vol. 50, no. 3, pp. 594–611, 2002, doi: 10.1109/22.989945.

[3] R. R. Mishra, and A. K. Sharma, "A review of research trends in microwave processing of metal-based materials and opportunities in microwave metal casting," *Crit. Rev. Solid State Mater. Sci.*, vol. 41, no. 3, pp. 217–255, 2016, doi: 10.1080/10408436.2016.1142421.

[4] A. C. Metaxas, and R. J. Meredith, *Industrial Microwave Heating*, 1988, doi: 10.1049/PBPO004E.

[5] J. Tang, and F. P. Resurreccion, "Electromagnetic basis of microwave heating," *Dev. Packag. Prod. Use Microw. Ovens*, pp. 3–38e, 2009, doi: 10.1533/9781845696573.1.3.

[6] M. Mehdizadeh, "Single-mode microwave cavities for material processing and sensing," *Microwave/RF Appl. Probes Mater. Heating, Sensing, Plasma Gener.*, pp. 109–150, 2010, doi: 10.1016/b978-0-8155-1592-0.00004-1.

[7] S. Chandrasekaran, S. Ramanathan, and T. Basak, "Microwave material processing-A review," *AIChE J.*, vol. 58, no. 2, pp. 330–363, 2012, doi: 10.1002/aic.

[8] R. Meredith, "Engineers' handbook of industrial microwave heating," *Engineers' Handbook of Industrial Microwave Heating*, 1998, doi: 10.1049/pbpo025e.

[9] R. R. Mishra, and A. K. Sharma, "Microwave-material interaction phenomena: Heating mechanisms, challenges and opportunities in material processing," *Compos. Part A Appl. Sci. Manuf.*, vol. 81, pp. 78–97, 2016, doi: 10.1016/j.compositesa.2015.10.035.

[10] H. Zhang, and A. K. Datta, "Microwave power absorption in single- and multiple-item foods," *Food Bioprod. Process. Trans. Inst. Chem. Eng. Part C*, vol. 81, no. 3, pp. 257–265, 2003, doi: 10.1205/096030803322438027.

[11] R. Rosa, P. Veronesi, and C. Leonelli, "A review on combustion synthesis intensification by means of microwave energy," *Chem. Eng. Process. Process Intensif.*, vol. 71, pp. 2–18, 2013, doi: 10.1016/j.cep.2013.02.007.

[12] J. Sun, W. Wang, and Q. Yue, "Review on microwave-matter interaction fundamentals and efficient microwave-associated heating strategies," *Materials (Basel).*, vol. 9, no. 4, 2016, doi: 10.3390/ma9040231.

[13] D. E. Clark and W. H. Sutton, "Microwave processing of materials," *Annu. Rev. Mater. Sci.*, vol. 26, pp. 299–331, 1996, doi: 10.1002/9781444355321.ch15.

[14] D. El Khaled, N. Novas, J. A. Gazquez, and F. Manzano-Agugliaro, "Microwave dielectric heating: Applications on metals processing," *Renew. Sustain. Energy Rev.*, vol. 82, no. December 2016, pp. 2880–2892, 2018, doi: 10.1016/j.rser.2017.10.043.

[15] A. K. Sharma, and R. R. Mishra, "Role of particle size in microwave processing of metallic material systems," *Mater. Sci. Technol. (United Kingdom)*, vol. 34, no. 2, pp. 123–137, 2018, doi: 10.1080/02670836.2017.1412043.

[16] Y. V. Bykov, K. I. Rybakov, and V. E. Semenov, "High-temperature microwave processing of materials," *J. Phys. D Appl. Phys*, vol. 34, no. 5, pp. R55–R75, 2001, doi: 10.1016/0924-2244(96)81279-9.

[17] M. Biesuz, T. Saunders, D. Ke, M. J. Reece, C. Hu, and S. Grasso, "A review of electromagnetic processing of materials (EPM): Heating, sintering, joining and forming," *J. Mater. Sci. Technol.*, vol. 69, pp. 239–272, 2021, doi: 10.1016/j.jmst.2020.06.049.

[18] S. A. A. Alem *et al.*, "Microwave sintering of ceramic reinforced metal matrix composites and their properties: A review," *Mater. Manuf. Process.*, vol. 35, no. 3, pp. 303–327, 2020, doi: 10.1080/10426914.2020.1718698.

[19] C. Prakash, S. Singh, S. Ramakrishna, G. Królczyk, and C. H. Le, "Microwave sintering of porous Ti—Nb-HA composite with high strength and enhanced bioactivity for implant applications," *J. Alloys Compd.*, vol. 824, 2020, doi: 10.1016/j.jallcom.2020.153774.

[20] C. Prakash *et al.*, "Processing of Ti50Nb50-xHAx composites by rapid microwave sintering technique for biomedical applications," *J. Mater. Res. Technol.*, vol. 9, no. 1, pp. 242–252, 2020, doi: 10.1016/j.jmrt.2019.10.051.

[21] R. R. Mishra, and A. K. Sharma, "Microstructural characteristics and tensile properties of in-situ and ex- situ microwave casts of Al-7039 alloy," *Mater. Res. Express*, 2019, [Online]. Available: https://iopscience.iop.org/article/10.1088/2053-1583/abe778.

[22] R. R. Mishra, and A. K. Sharma, "On mechanism of in-situ microwave casting of aluminium alloy 7039 and cast microstructure," *Mater. Des.*, vol. 112, pp. 97–106, 2016, doi: 10.1016/j.matdes.2016.09.041.

[23] R. R. Mishra, and A. K. Sharma, "On melting characteristics of bulk Al-7039 alloy during in-situ microwave casting," *Appl. Therm. Eng.*, vol. 111, pp. 660–675, 2017, doi: 10.1016/j.applthermaleng.2016.09.122.

[24] R. R. Mishra, and A. K. Sharma, "Structure-property correlation in Al-Zn-Mg alloy cast developed through in-situ microwave casting," *Mater. Sci. Eng. A*, vol. 688, no. February, pp. 532–544, 2017, doi: 10.1016/j.msea.2017.02.021.

[25] R. R. Mishra, and A. K. Sharma, "Effect of susceptor and mold material on microstructure of in-situ microwave casts of Al-Zn-Mg alloy," *Mater. Des.*, vol. 131, pp. 428–440, 2017, doi: 10.1016/j.matdes.2017.06.038.

[26] S. Singh, D. Gupta, and V. Jain, "Novel microwave composite casting process: Theory, feasibility and characterization," *Mater. Des.*, vol. 111, pp. 51–59, 2016, doi: 10.1016/j.matdes.2016.08.071.

[27] S. Singh, D. Gupta, and V. Jain, "Processing of Ni-WC-8Co MMC casting through microwave melting," *Mater. Manuf. Process.*, vol. 33, no. 1, pp. 26–34, 2018, doi: 10.1080/10426914.2017.1291954.

[28] A. K. Sharma, M. S. Srinath, and P. Kumar, "Microwave joining of metallic materials," Indian Patent application No. 1994/Del/2009, 2009.

[29] M. S. Srinath, A. K. Sharma, and P. Kumar, "Investigation on microstructural and mechanical properties of microwave processed dissimilar joints," *J. Manuf. Process.*, vol. 13, no. 2, pp. 141–146, 2011, doi: 10.1016/j.jmapro.2011.03.001.

[30] S. Singh, P. Singh, D. Gupta, V. Jain, R. Kumar, and S. Kaushal, "Development and characterization of microwave processed cast iron joint," *Eng. Sci. Technol. an Int. J.*, vol. 22, no. 2, pp. 569–577, 2019, doi: 10.1016/j.jestch.2018.10.012.

[31] D. Gamit, R. R. Mishra, and A. K. Sharma, "Joining of mild steel pipes using microwave hybrid heating at 2.45 GHz and joint characterization," *Journal of Manufacturing Processes*, vol. 27, pp. 158–168, 2017.

[32] R. Samyal, A. K. Bagha, and R. Bedi, "Evaluation of modal characteristics of SS202-SS202 lap joint produced using selective microwave hybrid heating," *J. Manuf. Process.*, vol. 68, no. PB, pp. 1–13, 2021, doi: 10.1016/j.jmapro.2021.07.018.

[33] D. Gupta, and A. K. Sharma, "Microwave cladding: A new approach in surface engineering," *Journal of Manufacturing Processes*, vol. 16, no. 2, pp. 176–182, 2014.

[34] B. Singh, and S. Zafar, "Understanding time-temperature characteristics in microwave cladding," *Manufacturing Letters*, vol. 25, pp. 75–80, 2020.

[35] S. Kaushal, V. Sirohi, D. Gupta, H. Bhowmick, and S. Singh, "Processing and characterization of composite cladding through microwave heating on martensitic steel," *Proceedings of the Institution of Mechanical Engineers, Part L: Journal of Materials: Design and Applications*, vol. 232, no. 1, pp. 80–86, 2018.

[36] A. M. Hebbale, and M. S. Srinath, "Taguchi analysis on erosive wear behavior of cobalt based microwave cladding on stainless steel AISI-420," *Measurement*, vol. 99, pp. 98–107, 2017.

Chapter 2

Microwave Drilling

Methods, Applications and Challenges

Anurag Singh, Gaurav Kumar, Pranjal Gupta,
and Apurbba Kumar Sharma

Contents

2.1 Introduction

Several nontraditional machining processes have emerged in the last few decades to deal with the limitations of conventional drilling. Processes, such as laser beam machining and electron beam machining, show excellent precision and accuracy. But most of these processes exhibit common limitations such as high initial setup cost, high operational and maintenance cost, low material removal rate (MRR) and high specific energy consumption [1]. Moreover, few processes are material-specific; for example, electrical discharge machining (EDM), electrochemical machining (ECM) and chemical machining (CHM) are generally restricted to metallic parts [2]. These limitations have motivated researchers to look for simpler, cheaper, versatile and energy-efficient alternatives. Microwave drilling is one such process being explored.

DOI: 10.1201/9781003248743-2

Microwave drilling is a novel nontraditional machining process. Like other thermal energy–based nontraditional machining processes, microwave drilling also relies on localized thermal energy to cause drilling at the target spot. However, it takes microwave energy as input and converts it into localized thermal energy through a heating method. Without making undesirable noise or fumes, it can drill through any material, regardless of conductivity or melting temperature [3]. Microwave drilling has outperformed other nontraditional processes in terms of drilling time and energy savings [4]. Thus, microwave drilling has the potential to replace conventional drilling at the commercial level [5]. However, further studies are required to fine-tune the process to minimize defects and meet industrial applications.

2.2 Methods of Microwave Drilling

Based on heating methods, a few methods of microwave drilling are being explored. Each method is based on a different microwave–matter interaction phenomenon and employs customized tooling. A tooling element interacts directly with microwave, concentrating and/or converting it into localized thermal energy. The physics of conversion of microwave energy into thermal energy is yet to be explored well. Some better-known methods of microwave drilling are discussed in the following sections.

2.2.1 Thermal Runaway–Based Drilling

Microwave-induced thermal runaway (or runaway heating) is a challenging microwave–matter interaction phenomenon. Microwave absorption (or dielectric loss factor) in certain materials tends to increase with rising temperature; this induces a thermal runaway condition, with a rapid and uncontrollable rise in the temperature of material subjected to microwave radiation [6]. Mainly, certain ceramics and polymers are known to exhibit thermal runaway [7]. Thermal runaway presents the issue of localized hot spots (or uneven heating) during microwave processing if the magnitude of the electric field varies significantly along the dimension of the material. However, the same issue is desirable for localized heating.

Jerby and Dikhtiar introduced microwave drilling in 2000 by utilizing the phenomenon of thermal runaway [8]. They synthesized a microwave device, consisting basically of a microwave source, coaxial waveguide and movable central electrode, to intentionally trigger the thermal runaway in the target object (Figure 2.1). Microwaves of specific wavelength generated at microwave source were transported via a coaxial waveguide to the movable central electrode, which was close to or in contact with the object. The movable central electrode concentrated the microwaves within a small volume of the object. Concentrated microwaves rapidly generated localized hot spot

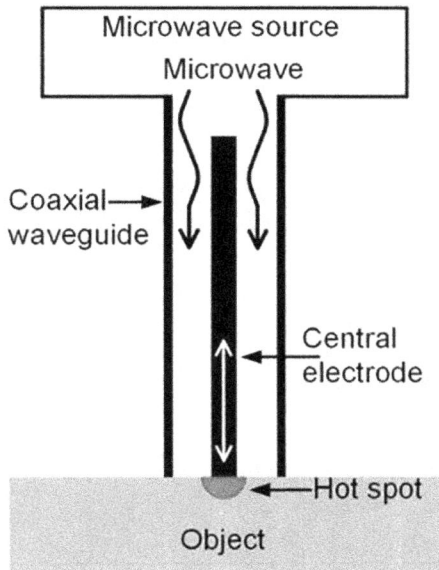

Figure 2.1 Principle of thermal runaway–based drilling.

by local thermal runaway. Subsequently, the movable central electrode was inserted into the softened volume to form the boundary of the hole.

The thermal runaway–based drilling method has been successfully employed to drill several hard (e.g., concrete, ceramics, glass and basalt) and soft materials (e.g., plastic) [8–21]. The process has shown the capability to drill micro-holes (< 1.0 mm), small holes (1.0–3.2 mm) and larger holes of various depths (Table 2.1). However, it is evident from Table 2.1 that thermal runaway–based drilling is limited to nonmetallic materials. It is because only certain nonmetallic materials can be locally heated by thermal runaway. Since metallic materials reflect incident microwaves, they cannot be heated locally to drill [11] [14]. Additionally, low-loss materials cannot be locally heated to drill [10]. However, this inherent limitation of material selectivity was advantageously utilized to drill ceramic thermal-barrier coating over the metallic substrate [18]. Thermal runaway–based drilling could be a quieter and cleaner alternative to noisy and dusty mechanical drilling due to the absence of fast rotating parts in the setup [8]. As of now, thermal runaway–based drilling cannot compete with laser drilling in terms of accuracy. However, it could be employed as a low-cost, quieter and cleaner alternative for general applications in the construction industry, geological surveys and industrial drilling systems [9].

Table 2.1 Experimental Conditions during Thermal Runaway–Based Drilling of Different Materials

Material	Power (W)	Hole Size (Ø × h) (mm²)	Drilling Time (s)	Issue
Concrete [8]	1000	5 × 30	~60	NA
Alumina [9]	900	6 × 13	>120	Molten debris around holes
Concrete [10]	1000	12 × 100	600–300	NA
Borosilicate glass [10]	NA	0.5 × ~1	NA	Inaccurate hole Required heat treatment to drill crack-less holes
Ceramic tiles [12]	100	1 × 5	~14	NA
Ertalon plastic [12]	25	1 × 9	~15	NA
Ovine tibias [13]	~200	2.4 × 3–5	<5	Carbonized margin around holes
Concrete [14]	600	~2 × ~20	<60	Glossy debris along hole
Concrete [17]	850	12 × 260	~2600	Resolidified debris inside and around hole
Soda-lime glass [20]	80	~1 × 1	~18	Needed slight grinding to remove porous debris to form the hole
Basalt [21]	600	2.4 × 20	~20	NA

Note: Ø = diameter of the drilled hole; h = thickness of sheet/plate; NA = not available.

Thermal runaway–based drilling method has considerably advanced in terms of setup, capability and efficiency since its inception. Auxiliaries such as isolator (to protect the microwave source against excessive microwave reflection), power monitor (to separate and measure microwave power forwarded to and reflected from the load) and E-H tuner (for impedance matching) were included in the basic laboratory setup, which was manually operated and controlled [14]. With the incorporation of real-time and automatic impedance matching and remote-controlled actuators, the setup has advanced from the laboratory stage [14] to the automatic and prototype stage [10] [15–17]. A basic thermal runaway–based drilling setup was limited to a drilling depth of $< \frac{\lambda}{4}$, where λ is the wavelength of microwave radiation [9]. Depth limitation was overcome by slowly rotating the coaxial waveguide as a hollow reamer and inserting it into the softened object [16]. The majority of thermal runaway–based drilling used magnetron-based microwave sources, which require high voltage power supplies, are rigid and are less sensitive to impedance mismatch. Whereas solid-state microwave sources are operable at low voltage supplies, frequency tunable and relatively sensitive to impedance mismatch [12]. Furthermore, they might be smaller, lighter and easier to manage [19].

2.2.2 Selective Microwave Hybrid Heating–Based Drilling

The microwave hybrid heating (MHH) method is employed to process materials that do not easily couple to microwaves around room temperatures, for example, some ceramics and bulk metallic materials [22–23]. Microwave hybrid heating employs an external heating source (microwave susceptor or conventional heating source) to raise the temperature of the load to a state where it starts to couple microwaves directly. Beyond the critical state, the temperature rises through self-heating and heat transfer from the external heating source. The selection of a suitable external heating source and separator decides the efficacy of MHH. The separator separates the load and facilitates heat transfer from the external heating source. Microwave hybrid heating is used for processes requiring a full extent of microwave exposure, that is, microwave melting and casting bulk metallic materials [24]. Selective microwave hybrid heating (SMHH) is MHH implemented at a localized portion of the load. It is commonly used for microwave joining of bulk metallic materials [25].

George et al. employed SMHH to demonstrate the viability of microwave drilling of metallic materials [26]. They designed a setup inside the microwave cavity consisting mainly of a spring-loaded tungsten drill bit (diameter: 2 mm) and charcoal susceptor powder (Figure 2.2). The bolt at the top of the setup held the spring, which kept the drill bit in contact with the metallic sheet. Microwave-friendly materials were used to cover the spring and drill bit to minimize microwave reflection. Charcoal susceptor powder

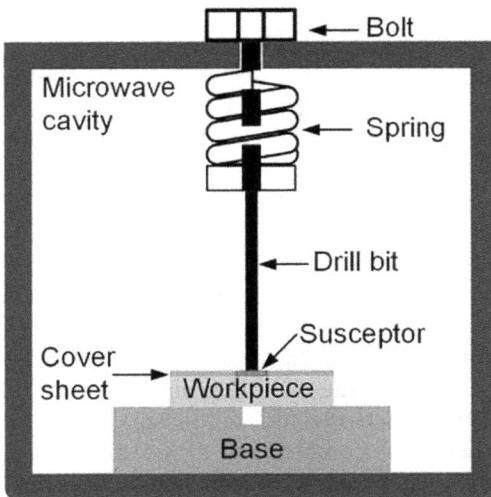

Figure 2.2 Selective microwave hybrid heating-based drilling setup.

was placed locally over the target spot on the metallic sheet. The rest of the sheet was covered by a concrete/graphite sheet to prevent microwave reflection. The charcoal susceptor directly coupled to microwaves and raised the temperature of the metallic sheet to a state where it started coupling to microwaves. Once softened enough due to heat transfer from the charcoal susceptor and self-heating, the spring-loaded tungsten drill bit penetrated the metallic sheet to form the hole. Sheets of 1.0-mm-thick aluminum, copper and stainless steel were drilled at a microwave power of 900 W. A hole was drilled in the aluminum sheet in 120 s, but the sheet partially burned due to overheating. Overheating was attributed to the low melting point of aluminum and the nonuniform placement of the charcoal susceptor powder at the target spot. A stainless steel sheet was only partially drilled even in drilling time of 240 s. However, the copper sheet was successfully drilled in a drilling time of 150 s. Another aluminum sheet of the same thickness was drilled in 60 s with a stainless steel drill bit with a 0.8-mm diameter [27]. The quality of the hole in terms of shape and finish was comparatively better.

The SMHH-based drilling method is tricky due to requirement of various tooling elements, such as bolt, beam, spring, drill bit, microwave-friendly materials, charcoal susceptor, and cover sheets. Because of the complexity involved in operating the method, fewer studies have been conducted.

2.2.3 Microwave–Metal Discharge–Based Drilling

Microwave-metal (MW-m) discharge (or microwave-induced discharge or microwave arcing) is a complex and inadequately understood microwave–matter interaction phenomenon. It is fundamentally an electrical discharge triggered when microwaves interact with the sharp points/edges of metallic or some semiconductive or carbon-based materials [28]. Microwave-metal discharge also exhibits specialized effects, namely, heating effect, plasma effect and photocatalytic effect. Despite the complexity, specialized effects of MW-m discharge are being used for various materials processing applications, such as reduction, pyrolysis, cracking, chemical synthesis and many more [29–32].

George et al. introduced MW-m discharge–based drilling by modifying the SMHH-based drilling setup [27]. They removed the concrete/graphite cover to drill nonmetallic materials as nonmetals do not reflect microwaves. Moreover, they removed the charcoal susceptor and spring load on the tungsten drill bit (diameter: 2 mm). In the modified setup, the function of the drill bit was to concentrate the incoming microwave at the tip of the drill bit and trigger MW-m discharge–induced plasma to generate a hot spot on the nonmetallic sheet. The combination of localized heat generated at the tip and the drill bit's self-weight was sufficient to drill through borosilicate glass and waste animal bone. Borosilicate glass of thickness 1.5 mm and 4.0 mm were drilled in 3 s and 6 s, respectively, at 700-W microwave power.

Most glasses broke due to the sudden application of heat and self-weight of the drill bit. Drilled holes exhibited cracks and poor surface finish. When exposed to 600 W of microwave power for 10 s, the MW-m discharge–induced plasma successfully drilled the bone (thickness: 6.0 mm) but burned it, resulting in a wide heat-affected zone (HAZ) and poor surface finish.

Das and Sharma developed a unique setup (coaxial cable applicator) to perform both SMHH-based drilling (of metallic materials) and MW-m discharge–based drilling (of nonmetallic materials) inside a metallic box with minor modifications [33]. Microwave generated from microwave source was transmitted via waveguide launcher to coaxial adapter, which finally transmitted microwaves via coaxial cable to monopole antenna in the metallic box. A monopole antenna of copper acted as a microwave concentrator and a drill bit. Selective microwave hybrid heating-based drilling of the aluminum sheets of thickness 1.0 mm was tried at a microwave power of 700 W. The aluminum sheet was covered at the drilling site with charcoal powder to prevent microwave reflection. Small MW-m discharge–induced spark formed at the contact point of the monopole antenna and the aluminum sheet. The coaxial cable burned after 5 s, leaving only an indentation imprint on the aluminum sheet. MW-m discharge–based drilling of borosilicate glass, waste animal bone and wood were also tried. The hole was drilled in borosilicate glass in 8 s with 360 W of microwave power; however, there was severe cracking. Wet bone (thickness: 6 mm) and wet wood (thickness: 5 mm) drilled at a microwave power of 500 W and 700 W, respectively, showed significantly lower HAZ than dry bone and dry wood.

The total efficiency of the coaxial cable applicator was substantially lower than microwave drilling inside the domestic microwave oven because of transmission loss in several long components [27]. The coaxial cable applicator was expected to drill both the metallic and nonmetallic materials, but it failed to drill aluminum sheets. The experimentation duration was restricted to about 10 s due to the inability of the coaxial cable to sustain high temperatures [27]. Using a coaxial waveguide instead of a coaxial cable might extend the duration of the experiment and allow the drilling of metallic materials. On the other hand, the coaxial waveguide will limit the spatial flexibility of the metallic box (or applicator).

MW-m discharge–based drilling (within the domestic microwave oven) has mostly been used to drill borosilicate glass, Perspex and soda-lime glass since its inception. Better tooling or/and process parameter optimization are improving the quality of drilled holes, which can be seen in Table 2.2. However, the technique still needs to be fine-tuned as the quality is still subpar.

Recently, Singh and Sharma demonstrated the drilling of metallic materials using MW-m discharge [41]. They designed a simplified setup inside the domestic microwave oven that consisted primarily of a pin vise, pointed thoriated tungsten tool (diameter: 1.0 mm) and refractory base (Figure 2.3a). Metallic pin vise gripped the tool over the metallic sheet under a preset tool–work gap

Table 2.2 Effect of Tooling and Parameter Optimization on Quality of MW-m Discharge Drilled Holes in Nonmetallic Materials

Material	Tooling	Parameter	Effect on Quality	Year
Borosilicate glass	Gravity-fed tungsten rod tool is in contact with glass in atmospheric air.	Power of 700 W for 3–6 s.	Thermal shock from plasma broke most glasses and led to cracks and poor surface finish.	2012 [27]
	Gravity-fed cylindrical tool is in contact with glass in dielectrics.	Power of 700 W for 4s.	Dielectrics confined plasma and reduced thermal shock, resulting in diminished cracks.	2018 [34]
	Tool feeding system fed conical tool with gap and preheated glass is in dielectrics.	Power of 700 W for 3–12 s.	Feeding system synced ablation rate with feed rate. Gap allowed debris flushing. Preheating further reduced thermal shock. All resulted in less thermal damage.	2020 [3]
	Gravity-fed conical tool is in contact with glass in flowing dielectrics.	Power of 70–700 W for 10–50 s	Dielectrics' flow enhanced heat dissipation and debris flushing, resulting in less damage.	2020 [4]
Perspex	Gravity-fed tool is in contact with Perspex in atmospheric air.	Power of 700 W for 60 s	Poor circularity, significant overcut and burrs and wide heat-affected zone (HAZ).	2014 [35]
		Power of 90–900 W for 4–50 s	Better circularity, reduced overcut and narrow HAZ at lower power and drilling time.	2019 [36–37]
Soda-lime glass	Gravity-fed tool is in contact with glass in atmospheric air.	Power of 90–900 W for 30–60 s	Plasma interacting with glass caused thermal stresses in the HAZ, which led to cracking.	2015 [38]
	Precursor surrounds gravity-fed tool contacting the glass.	Power of 90–900 W for ≤4 s	Precursors aided in the reduction of thermal shock, which in turn lowered the likelihood of cracking.	2015 [39]
	Gravity-fed tool is in contact with glass in atmospheric air.	Power of 90–900 W for ≤4 s	90 W for 4 s yields better results due to localized heating owing to low-volume plasma.	2015 [40]

(1–2 mm). The pin vise was either held vertical by a microwave transparent holder supported on two insulating supports or by a metallic holder at the top of the microwave oven. A low microwave-absorbing refractory base supported the metallic sheet. They performed MW-m discharge–based drilling for 60–70 s at 700 W of microwave power. They demonstrated that the temperature of the MW-m discharge–induced plasma channel was high enough to drill stainless steel sheets. They also recorded and analyzed the MW-m discharge-based drilling [42]. They observed that the transverse diameter of the MW-m discharge-induced plasma channel was significantly larger than the tool diameter, and it varied with time while the shape remained constant.

Furthermore, based on the experimental results of MW-m discharge–based drilling of galvanized steel and stainless steel sheets at 700 W of microwave power (Figure 2.3b and c), they hypothesized the mechanism of formation of MW-m discharge-induced plasma channel in atmospheric air [43]. Interaction of microwave with the pointed thoriated tungsten tool resulted in a denser concentration of free electrons at the pointed end than at the cylindrical length of the tool, intensifying the electric field around the tool tip. It was proposed that free electrons traveled rapidly through the tool–work gap under the influence of the intensified electric field. The rapid traverse of free electrons led to ionization, electrical breakdown and, ultimately, the formation of MW-m discharge–induced plasma channel. Since the tool–work gap is least at the pointed tip, MW-m discharge–induced plasma channel is localized around the tip. Under similar experimental settings, MW-m discharge–based drilling was recently used to drill copper sheets (thickness: 0.4 mm) and Inconel sheets (thickness: 0.6 mm; Figure 2.3d and e). However, geometrical inaccuracies, thermal damage and poor process repeatability were observed in all cases of MW-m discharge–based drilling of metallic materials. It could be attributed to a lack of control over the MW-m discharge–induced plasma channel. Recent reports on MW-m discharge–based drilling of metallic materials prove the versatility of the MW-m discharge–based drilling method. However, further developments in tooling design and parameter optimization are required to advance the process.

Each method of microwave drilling differs from another considerably. Table 2.3, which lists the characteristics of each microwave drilling method, demonstrates this. The thermal runaway–based drilling method is clearly limited to lossy dielectrics. Similarly, the SMHH-based drilled method is limited to metallic materials. On the other hand, the MW-m discharge–based drilling technology can drill metallic and nonmetallic materials.

2.3 Challenges of Microwave Drilling Methods

Microwave drilling has been studied as an alternative to established non-traditional machining technologies for the past two decades. Several encouraging results were obtained with each method of microwave drilling throughout

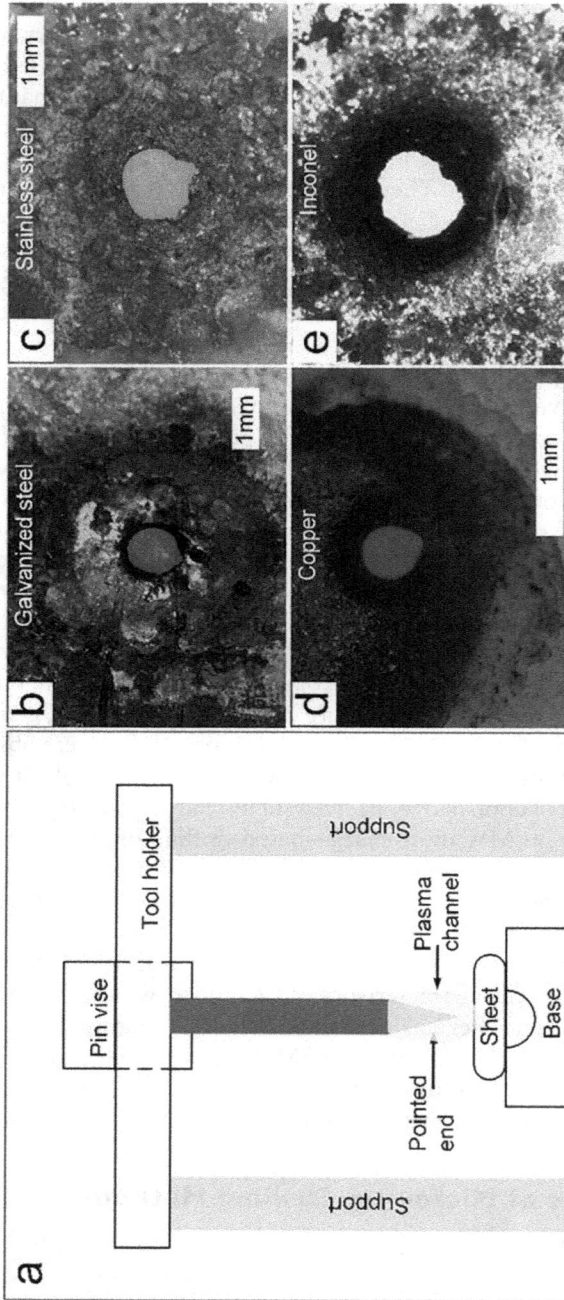

Figure 2.3 (a) Experimental setup for MW-m discharge–based drilling of metallic materials (after [43]). Drilling in (b) galvanized (c) stainless steel (d) copper and (e) Inconel sheet

Table 2.3 Comparison of Microwave Drilling Methods

Methods Attribute	Thermal Runaway– Based Drilling	SMHH-Based Drilling	MW-m Discharge– Based Drilling
Tool–work contact	Yes	Yes	Not necessarily
Suitability	Lossy dielectrics	Metallic materials	Nonmetallic and metallic materials
Material removal	Softening followed by drill insertion	Softening followed by drill insertion	Melting and/or vaporization
MW leakage	Yes but reduced to common safety standards	No	No
Developmental stage	Prototype	Laboratory	Laboratory
Potential	Silent and dust-free alternative to the mechanical drill	–	Mono-electrode discharge–based drilling method

this period. However, microwave drilling is in its early stages of development and is not yet ready for industrial use. Each approach has unique challenges that must be overcome to advance the process. In the following sections, the challenges associated with each method of microwave drilling are briefly discussed.

2.3.1 Challenges of Thermal Runaway–Based Drilling

Thermal runaway–based drilling method (microwave drill) has evolved from a basic stage restricted to drilling shallow holes to an advanced prototype capable of drilling deeper holes [16] [17]. Drilling depth with the basic setup is obtained by mechanically inserting the concentrator's central electrode into the softened hot spot. However, the insertion depth of the central electrode is limited to $\lambda/4$, which, in turn, limits the drilling depth. Drilling depth was extended by incorporating two more simultaneous motions in the advanced prototype in addition to the central electrode's axial insertion, slow rotation of coaxial outer cylinder/tube and mechanical insertion of the entire concentrator independent of the electrode's axial insertion. The advanced prototype also incorporates adaptive impedance matching and remote control. An adaptive impedance matches the lower microwave reflection from the load and adjusts the microwave power according to the variation in the load impedance and thus help improve process efficiency. During the synchronized operation of the advanced prototype, the central electrode is initially inserted into the hot spot to the depth of $\lambda/4$. Simultaneously, the rotating tube (with helical grooves on its outside) functioning as a hollow reamer cuts through the softened zone to widen the hole and perform debris evacuation assisted by pressurized airflow through the hollow tube. After debris removal, the entire concentrator is axially inserted into the widened hole to deepen the hole.

The advanced prototype ensured deeper hole drilling, but the drilling speed was slow (<10mm/min) considering practical applications. The challenge is raising the drilling speed to at least five times of current value. Enhancement in overall drilling speed could be achieved by improving debris removal and expediting the response of adaptive impedance matching.

2.3.2 Challenges of SMHH–Based Drilling

The selective microwave hybrid heating–based drilling method established the microwave drilling of metallic materials in 2012 [26]. Since then, this method has not been explored as other methods of microwave drilling. This lack of exploration may be attributed to tooling challenges observed while implementing SMHH-based drilling. The foremost challenge is the selection of microwave susceptor and its uniform distribution of optimum amount at the drilling site. An excess of the microwave susceptor may result in thermal damage in the zone near the drilled hole, whereas a lesser microwave susceptor may fail to raise the temperature of the drilling site enough to be penetrated by spring load. Spring load must be calculated considering geometrical attributes and mechanical properties of the material to be drilled and expected temperature rise by SMHH. Excessive microwave reflection from the spring-loaded metallic drill bit and metallic workpiece may decrease the process efficiency; the magnetron may also get damaged. Thus, a reasonable selection of materials for covering the spring-loaded drill bit and the metallic sheet is another challenge. These materials are required to be an absorber of microwaves to prevent excessive microwave reflection but not too absorptive to consume most of the incoming microwave energy.

2.3.3 Challenges of MW-m Discharge-Based Drilling

MW-m discharge–based drilling method presents numerous challenges due to the lack of understanding of the physics of MW-m discharge, which is still inadequately explored. Moreover, its dependence on numerous factors adds to complexity and processing challenges. Some of these factors are amenable, such as microwave power, microwave exposure duration and medium. However, the geometry and morphology of the tool, which is primarily responsible for the erratic nature of MW-m discharge, are very difficult to manage. Furthermore, the inability to access real-time data of temperature, electron density and the like of the MW-m discharge–induced plasma channel during the drilling makes it difficult to develop its mathematical model and explain the physics. Following are the major challenges in the MW-m discharge–based drilling method.

2.3.3.1 Generation of Continuous and Stable MW-m Discharge–Induced Plasma Channel

The continuity and stability of Mw-m discharge–induced plasma channel mainly depend on the geometry of sharp/conical tool. However, the geometry of the tool may change during the process due to thermal erosion by a high-temperature plasma channel. A change in the geometry of the tool may make MW-m discharge-induced plasma channel discontinuous/unstable or stop altogether. Thus, the geometry must be maintained to trigger a continuous and stable MW-m discharge–induced plasma channel. Geometry can be maintained by using high-strength, high-temperature-resistant and high-thermal-conductivity materials for tool design. High strength and temperature resistance will impart thermal stability against the high-temperature plasma channel. A high thermal conductivity will help dissipate thermal energy away from the sharp point.

2.3.3.2 Geometrical and Thermal Characterization of MW-m Discharge–Induced Plasma Channel

A characterization of the MW-m discharge–induced plasma channel will help with understanding the nature and processing capability of the MW-m discharge–based drilling method. Geometrical characterization requires a high-speed camera with reasonable resolution. Similarly, the thermal characterization needs a high-speed thermal imaging device with excellent spatial and temporal resolution.

2.3.3.3 Reduction of Geometrical Inaccuracies and Thermal Damage

MW-m discharge–based drilling method is capable of drilling both nonmetallic and metallic materials. But the extent of geometrical inaccuracies and thermal damage observed is significant. Like other processes, the ultimate challenge of MW-m discharge–based drilling is minimizing geometrical and thermal defects, which could be attained with better control over the characteristics of the MW-m discharge–induced plasma channel.

2.4 Conclusion

Microwave drilling is an emerging nontraditional machining process. The principles, capabilities, limitations and challenges of methods of microwave drilling were discussed in this chapter. Unlike established nontraditional machining processes that are only suitable for certain types of materials, microwave drilling can drill various solid materials, including metals, ceramics, polymers and composites. Microwave drilling performed through different methods have different tooling, capabilities, limitations and challenges. Drilling deeper holes in concrete has been achieved with the thermal runaway–based drilling approach, best suited for lossy dielectrics.

However, drilling speed is slow, which can be improved with better debris evacuation and faster adaptive impedance matching. Selective microwave–hybrid heating–based drilling method is tricky and restricted to metallic materials. MW-m discharge–based drilling method can drill both nonmetallic and metallic materials. However, it is facing numerous challenges due to a lack of control over the MW-m discharge–induced plasma channel. Microwave drilling is still in its early stages of development, and it will take a lot of research and development to bring it up to industry standards.

References

[1] Jain, N.K., Jain, V.K. and Deb, K., 2007. Optimization of process parameters of mechanical type advanced machining processes using genetic algorithms. *International Journal of Machine Tools and Manufacture, 47*(6), pp. 900–919.

[2] Groover, M.P., 2020. *Fundamentals of Modern Manufacturing: Materials, Processes, and Systems.* John Wiley & Sons.

[3] Kumar, G. and Sharma, A.K., 2020. On processing strategy to minimize defects while drilling borosilicate glass with microwave energy. *The International Journal of Advanced Manufacturing Technology, 108*(11), pp. 3517–3536.

[4] Kumar, G., Mishra, R.R. and Sharma, A.K., 2021. On defect minimization during microwave drilling of borosilicate glass at 2.45 GHz using flowing dielectric and optimized input power. *Journal of Thermal Science and Engineering Applications, 13*(3), p. 031021.

[5] Loharkar, P.K., Ingle, A. and Jhavar, S., 2019. Parametric review of microwave-based materials processing and its applications. *Journal of Materials Research and Technology, 8*(3), pp. 3306–3326.

[6] Horikoshi, S., Schiffmann, R.F., Fukushima, J. and Serpone, N., 2018. *Microwave Chemical and Materials Processing* (pp. 1–17). Springer, Singapore.

[7] Santos, T., Valente, M.A., Monteiro, J., Sousa, J. and Costa, L.C., 2011. Electromagnetic and thermal history during microwave heating. *Applied Thermal Engineering, 31*(16), pp. 3255–3261.

[8] Jerby, E. and Dikhtiar, V., Ramot at Tel Aviv University Ltd, 2000. *Method and Device for Drilling, Cutting, Nailing and Joining Solid Non-Conductive Materials Using Microwave Radiation.* U.S. Patent 6,114,676.

[9] Jerby, E., Dikhtyar, V., Aktushev, O. and Grosglick, U., 2002. The microwave drill. *Science, 298*(5593), pp. 587–589.

[10] Jerby, E., Aktushev, O., Dikhtyar, V., Livshits, P., Anaton, A., Yacoby, T., Flax, A., Inberg, A. and Armoni, D., 2004 November. Microwave drill applications for concrete, glass and silicon. In *Proc. 4th World Congress Microw. Radio-Frequency Applications* (pp. 4–7). American Institute of Chemical Engineers, New York.

[11] Jerby, E., Dikhtyar, V. and Aktushev, O., 2003. Microwave drill for ceramics. *American Ceramic Society Bulletin, 82*(1), pp. 35–37.

[12] Mela, O. and Jerby, E., 2008 March. Miniature transistor-based microwave drill. In *Proc. Global Congress Microwave Energy Applications* (pp. 443–446). Otsu, Japan.

[13] Eshet, Y., Mann, R.R., Anaton, A., Yacoby, T., Gefen, A. and Jerby, E., 2006. Microwave drilling of bones. *IEEE Transactions on Biomedical Engineering*, *53*(6), pp. 1174–1182.

[14] Jerby, E. and Dikhtyar, V., 2006. Drilling into hard non-conductive materials by localized microwave radiation. In *Advances in Microwave and Radio Frequency Processing* (pp. 687–694). Springer, Berlin, Heidelberg.

[15] Jerby, E., Dikhtyar, V. and Aktushev, O., 2004 September. The microwave-drill technology. In *2004 23rd IEEE Convention of Electrical and Electronics Engineers in Israel* (pp. 269–272). IEEE, Tel-Aviv, Israel.

[16] Jerby, E., Shamir, Y., Peleg, R. and Aharoni, Y., 2013 September. A silent mechanically-assisted microwave-drill for concrete with integrated adaptive impedance matching. In *Proc. AMPERE Int. Conf. Microw. RF Heatin* (pp. 267–270). Nottingham, England.

[17] Jerby, E., Nerovny, Y., Meir, Y., Korin, O., Peleg, R. and Shamir, Y., 2017. A silent microwave drill for deep holes in concrete. *IEEE Transactions on Microwave Theory and Techniques*, *66*(1), pp. 522–529.

[18] Jerby, E. and Thompson, A.M., 2004. Microwave drilling of ceramic thermal-barrier coatings. *Journal of the American Ceramic Society*, *87*(2), pp. 308–310.

[19] Meir, Y. and Jerby, E., 2011 November. Transistor-based miniature microwave-drill applicator. In *2011 IEEE International Conference on Microwaves, Communications, Antennas and Electronic Systems (COMCAS 2011)* (pp. 1–4). IEEE, New York.

[20] Meir, Y. and Jerby, E., 2012. Localized rapid heating by low-power solid-state microwave drill. *IEEE Transactions on Microwave Theory and Techniques*, *60*(8), pp. 2665–2672.

[21] Jerby, E., Dikhtyar, V. and Einat, M., 2004 November. Microwave melting and drilling of basalt. In *Proc. AIChE Annu. Meeting* (pp. 4–7). Nottingham, England.

[22] Gupta, M. and Leong, E.W.W., 2008. *Microwaves and metals*. John Wiley & Sons.

[23] Tiwari, A., Gerhardt, R.A. and Szutkowska, M. eds., 2016. *Advanced ceramic materials*. John Wiley & Sons.

[24] Sharma, A.K. and Mishra, R.R., 2018. Challenges in Microwave Processing of Bulk Metallic Materials and Recent Developments. *AMPERE Newsletter* (ISSN 1361-8598), Issue 96.

[25] Mishra, R.R. and Sharma, A.K., 2016. A review of research trends in microwave processing of metal-based materials and opportunities in microwave metal casting. *Critical Reviews in Solid State and Materials Sciences*, *41*(3), pp. 217–255.

[26] George, T.J., Sharma, A.K. and Kumar, P., 2012. A feasibility study on drilling of metals through microwave heating. *i-Manager's Journal on Mechanical Engineering*, *2*(2), pp. 1–6.

[27] George, T.J., Sharma, A.K., Kumar, P., Kumar, R. and Das, S., 2012 September. Microwave drilling: Future possibilities and challenges based on experimental studies. In *International Conference on Emerging Trends in Manufacturing Technology, Toc H Institute of Science & Technology*, Kerela.

[28] Sun, J., Wang, W., Yue, Q., Ma, C., Zhang, J., Zhao, X. and Song, Z., 2016. Review on microwave: Metal discharges and their applications in energy and industrial processes. *Applied Energy*, *175*, pp. 141–157.

[29] Standish, N. and Worner, H., 1990. Microwave application in the reduction of metal oxides with carbon. *Journal of Microwave Power and Electromagnetic Energy*, 25(3), pp. 177–180.

[30] Sun, J., Wang, W., Liu, Z. and Ma, C., 2011. Recycling of waste printed circuit boards by microwave-induced pyrolysis and featured mechanical processing. *Industrial & Engineering Chemistry Research*, 50(20), pp. 11763–11769.

[31] Sun, J., Wang, Q., Wang, W., Song, Z., Zhao, X., Mao, Y. and Ma, C., 2017. Novel treatment of a biomass tar model compound via microwave-metal discharges. *Fuel*, 207, pp. 121–125.

[32] Sun, J., Wang, Y., Wang, W., Wang, K. and Lu, J., 2018. Application of featured microwave-metal discharge for the fabrication of well-graphitized carbon-encapsulated Fe nanoparticles for enhancing microwave absorption efficiency. *Fuel*, 233, pp. 669–676.

[33] Das, S. and Sharma, A.K., 2012. Microwave drilling of materials. *BARC Newslett*, 329, pp. 15–21.

[34] Kumar, G. and Sharma, A.K., 2018. Role of dielectric fluid and concentrator material in microwave drilling of borosilicate glass. *Journal of Manufacturing Processes*, 33, pp. 184–193.

[35] Lautre, N.K., Sharma, A.K. and Kumar, P., 2014 June. Distortions in hole and tool during microwave drilling of perspex in a customized applicator. In *2014 IEEE MTT-S International Microwave Symposium (IMS2014)* (pp. 1–3). IEEE, New York.

[36] Lautre, N.K., Sharma, A.K., Kumar, P. and Das, S., 2019. Characterization of drilled hole in low melting point material during low power microwave drilling process. *Materials Research Express*, 6(9), p. 095329.

[37] Lautre, N.K., Sharma, A.K., Kumar, P. and Das, S., 2018. Experimental evaluation of a microwave drilling process in perspex. *Journal of Testing and Evaluation*, 48(4), pp. 2880–2894.

[38] Lautre, N.K., Sharma, A.K., Kumar, P. and Das, S., 2015. A photoelasticity approach for characterization of defects in microwave drilling of soda lime glass. *Journal of Materials Processing Technology*, 225, pp. 151–161.

[39] Lautre, N.K., Sharma, A.K., Das, S. and Kumar, P., 2015. On crack control strategy in near-field microwave drilling of soda lime glass using precursors. *Journal of Thermal Science and Engineering Applications*, 7(4).

[40] Lautre, N.K., Sharma, A.K., Pradeep, K. and Das, S., 2015. A simulation approach to material removal in microwave drilling of soda lime glass at 2.45 GHz. *Applied Physics A*, 120(4), pp. 1261–1274.

[41] Singh, A. and Sharma, A.K., 2019. Temperature profiling of microwave: Metal discharge plasma channel using image processing technique. In *Advances in Forming, Machining and Automation* (pp. 219–227). Springer, Singapore.

[42] Singh, A. and Sharma, A.K., 2019. Effect of drilling conditions on microwave-metal discharge during microwave drilling of stainless steel. In *Applied Mechanics and Materials* (Vol. 895, pp. 253–258). Trans Tech Publications Ltd., Stafa-Zurich.

[43] Singh, A. and Sharma, A.K., 2020. On microwave drilling of metal-based materials at 2.45 GHz. *Applied Physics A*, 126(10), pp. 1–11.

Chapter 3

Microwave Drilling in Sub-Wavelength Diameters

Eli Jerby

Contents

3.1 Introduction

Drilling holes is a fundamental operation in industrial, constructional, geological, and medical processes [1–9]. Mechanical drills satisfy many of these needs, but their operation causes noise, vibrations, and dust effusion. Advanced drilling methods based on thermal and ablation effects make use of lasers, plasma jets, flames, and electrical heaters [2, 3]. Other methods employ ultrasonic devices, water jets, hydraulic presses, electron beams, and

DOI: 10.1201/9781003248743-3

electro-erosion tools [4–6]. This chapter reviews another drilling method, based on the localized microwave-heating (LMH) effect in sub-wavelength diameters [10], and summarizes a two-decade research-and-development activity on this topic.

Microwaves are widely used in communication and radar systems, and they also have great potential (although not fully utilized yet) in industrial, scientific, and medical (ISM) applications [11, 12]. Current industrial applications of microwaves include heating, drying, material processing and manufacturing, ceramic sintering, and microwave chemistry [13, 14]. Applications of microwave energy for destructive purposes, such crushing rocks for mining operations, tunneling, and so on [7, 15], as well as demolition of concrete structures, have also been proposed and developed but typically for larger size scales compared to the microwave wavelength λ (~0.12 m at 2.45 GHz, or ~0.33 m at 915 MHz). The heat-affected zone (HAZ) size in microwave heating is typically in the order of the microwave wavelength, which is approximately 10^4 longer than that of a CO_2 laser, for instance. Hence, a remote, far-field focusing of microwaves for fine drilling operations, such as performed by laser-based drills [16], is limited to the minimal order of the microwave wavelength, due to the diffraction limit.

Significant difficulties in large-scale ISM applications are caused by the thermal-runaway instability [17–23], in which nonuniform microwave heating may accidentally cause a rapid increase of the local temperature, thus creating a harmful hot spot that may severely damage the processed material (in particular in microwave-heating processes that require uniformity, such as sintering or drying).

On the contrary, the thermal-runaway effect, undesired in most applications, is purposefully used in the microwave drilling concept [10, 24–31]. It enables a specifically designed, near-field microwave applicator, namely, the *microwave drill*, to concentrate the microwave-heating energy into a hot spot much smaller than the microwave wavelength, and hence to locally soften and melt the confined drilled volume. Using the near-field LMH effect, experimental studies demonstrate the ability of microwave drills to make relatively delicate holes and cuts, down to about an 0.5-mm diameter in glass (more than two orders of magnitude smaller than the microwave wavelength applied). Microwave drills have also been studied in a variety of other materials [26, 27], such as concrete, ceramics [28, 29], basalts [30, 32], silicon, silicate minerals, bones [31], Perspex [33], and thin metal sheets [34]. A solid-state microwave drill with approximately 0.1 kW of power [35] has demonstrated a stepwise microwave drilling of about 4-mm-deep holes in glass [36].

In a wider scope, similar near-field microwave-heating effects induced by open-end applicators were studied for various applications, including silicon heating [37] and doping [38], electrochemical processes in liquids [39–42], interstitial treatment [43, 44], tissue heating [45], and tissue ablation for cancer treatment [46]. The LMH paradigm [47, 48], using microwave-drill-like applicators, has been extended to other applications, such as cutting, joining [49], additive manufacturing, combustion, plasma ejection [50], and nano-powder generation [51]. This relatively new field of sub-wavelength microwave heating has potential applications in various industries (e.g., electronics [52], ceramics, automotive, avionics and space), as well as in geological and construction work.

Most of the microwave drilling capabilities presented in the literature were first demonstrated using magnetron-based systems (at up to ~1 kW regulated power), as used for instance in domestic microwave ovens. These systems require 4-kV power supplies, waveguide sections, and impedance matching elements, which are too cumbersome for portable machines or delicate operations. Hence, using compact solid-state microwave sources (instead of magnetrons) reduces the size, weight, and operating voltage and improves the controllability, tunability and spectral characteristics of the microwave drills. A similar transition from vacuum to solid-state electronics was proposed decades ago for volumetric (non-LMH) microwave heating [53, 54] though yet (in 2021) most of the operating microwave-heating systems still employ magnetrons due to their other advantages.

The next sections review the LMH principles and theory behind the microwave drilling mechanism, present various examples of microwave drill experiments and results related to applications, and discuss the technological potential and outlook.

3.2 Sub-Wavelength Microwave Drill Principles

A conceptual illustration of a basic microwave drill applicator and an early prototype of it are shown in Figures 3.1a and 3.1b, respectively. This open-end applicator consists of a coaxial structure with an extendable center electrode, which also functions as a drill bit. Due to the LMH effect, the near-field microwave power tends to concentrate into a relatively small volume in the drilled material, in front of the applicator [55]. The absorbed microwave energy is dissipated then in this confined region and locally heats it up. The thermal-runaway instability is attributed to the tendency of the hotter region to better absorb microwaves than its cooler vicinity. This self-localized (LMH) effect rapidly evolves to a hot spot, as shown in Figure 3.1c.

The LMH effect featured a positive nonlinear feedback loop, since the temperature rise is locally accelerated up to the melting point. Additional microwave irradiation may also cause evaporation and even plasma ejection from the hot spot [51]. Unlike the remote laser-based drill, the microwave drill's near-field HAZ is much smaller than the radiation wavelength. After the hot spot formation, the extendable center electrode of the coaxial applicator is inserted into the softened HAZ, in order to advance the drilling process and to form the hole. Finally, the electrode is pulled out from the drilled

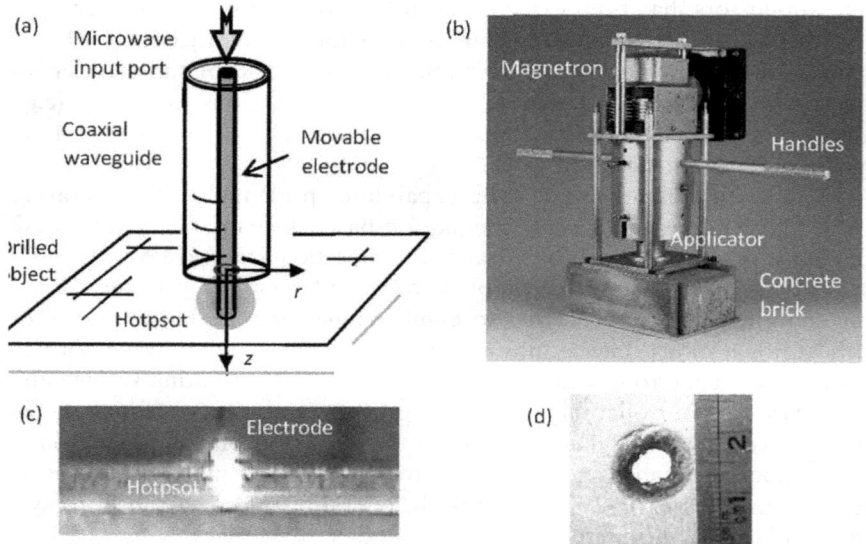

Figure 3.1 A basic microwave drill scheme: (a) A simplified illustration of the microwave drill applicator, consisting of a coaxial waveguide and a moveable center electrode. (b) An early experimental prototype of a manually operated microwave drill for concrete bricks. (c) A hot spot created in a glass plate by a 1-mm⌀ electrode. (d) A microwave-drilled hole in low-purity alumina, with the porous debris left on the bore perimeter in order to show the LMH effect [10].

hole, while the material cools down in its new shape. This basic microwave drill technique is applicable for a variety of non-conductive materials as listed above. The applicable materials should be susceptible to the thermal-runaway instability due to their temperature-dependent characteristics; that is, their dielectric-loss tends to increase (and/or thermal conductivity to decrease) as the temperature rises. The basic microwave drill does not require rotating parts, and it makes no dust or noise.

The basic scheme illustrated in Figure 3.1a is applicable, however, to limited ranges of hole dimensions and material properties. Its effective drilling depth is limited to ~λ/4 within the material (~2 cm in concrete at 2.45 GHz). In deeper penetration depths the broadside radiation is increased on the expense of the effective on-axis component, hence reducing the microwave drilling effectiveness. Figures 3.2a and 3.2b show for instance the microwave drill's center electrode inserted into a concrete brick (as in Figure 3.1b) and remained inside. As evidenced from the X-ray image in Figure 3.2b, this 3.2-mm⌀ electrode exceeds an approximately 2.5-cm insertion depth, in accord to the ~λ/4 penetration limit. Nails inserted by this technique resist extraction forces up to 30 kg in strength.

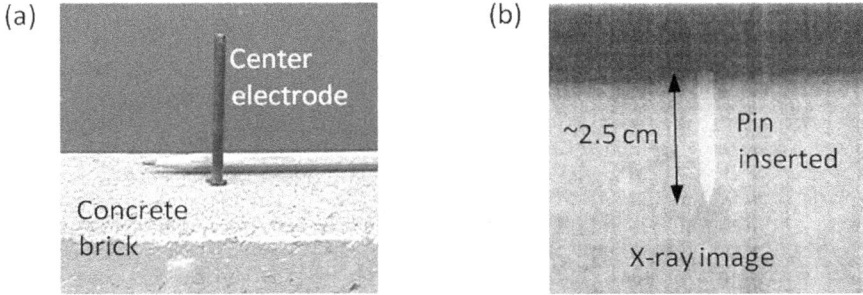

Figure 3.2 A 3.2-mm$^\varnothing$ center electrode of the microwave drill shown in Figure 3.1b, inserted into a concrete brick and remains as a nail inside: (a) An outer view of the nail inserted and (b) an X-ray image of the nail inside the brick, exceeding an approximate 2.5-cm insertion depth [36].

Advanced microwave drill schemes, as presented in Section 3.4, enable wider operating ranges, and demonstrate hole diameters in the range from about 0.5 mm in glass to 12 mm in concrete (the latter in >20-cm hole depths). The approximate $\lambda/4$ limit is alleviated for deeper and larger diameter holes, by adding to the outer cylinder a function of a hollow reamer, which is also penetrating into the hole, shaping and deepening it, while it is slowly rotating and removing the debris. Besides drilling, this LMH-based technology enables the insertion of pins and nails (as in Figure 3.2); melting [32], cutting [55] and jointing operations; and advanced applications such as additive manufacturing (3D-printing) [57], doping [38], plasma ejection [51, 58], ignition and combustion [59, 60].

3.3 LMH Model

This section presents a theoretical model for the intentional thermal-runaway process, applied by microwave drills in local sub-wavelength scales, as illustrated in Figure 3.1 [55, 61, 62]. The theoretical model couples the electromagnetic (EM) wave equation and the heat equation, taking into account the temperature dependencies of the material properties, as well as heat-convection and blackbody-radiation effects. The numerical LMH analysis is simplified by the *two time-scale* approximation [55], assuming that the EM-wave propagation is much faster than the thermal evolution. The coupled set of EM-heat equations consists of the wave equation in the frequency domain,

$$\nabla \times \left(\nabla \times \tilde{\mathbf{E}} \right) - \left[\varepsilon_r' - j \left(\varepsilon_r'' + \sigma_c / \omega \varepsilon_0 \right) \right] k_0^2 \tilde{\mathbf{E}} = 0, \qquad (3.1)$$

and the heat equation in the time domain [63],

$$\rho c_p \, \partial T/\partial t - \nabla \cdot (k_{th} \nabla T) = Q_d, \tag{3.2}$$

coupled by the dissipated power density,

$$Q_d = \frac{1}{2}(\omega \varepsilon_0 \varepsilon_r'' + \sigma_c)|\tilde{\mathbf{E}}|^2, \tag{3.3}$$

where $\tilde{\mathbf{E}}$, k_0, and ω are the electric-field vector, the wavenumber and angular frequency, respectively, of the EM-wave phasor in the frequency domain. The drilled material is represented by $\varepsilon(T) = \varepsilon_0(\varepsilon_r' - j\varepsilon_r'')$ and σ_c, the complex dielectric-permittivity and electric conductivity, respectively. In the heat equation (Equation 3.2), ρ, c_p, and k_{th}, are the local density, heat capacity and thermal conductivity, respectively, and $T(t)$ is the slowly varying temperature in the time domain. The material parameters, σ_c, ρ, c_p and k_{th}, are considered as temperature-dependent variables (as in $\varepsilon(T)$). Hence, Equations 3.1 through 3.3 are also coupled by the temperature dependencies of the material parameters, as the temperature profile is dynamically evolved in time and space.

The *two time-scale* approximation allows the solution of the EM-wave (Equation 3.1) in the frequency domain and the heat equation 3.2 in the time domain, due to the distinction between the typical time scales of the EM wave propagation and the much slower thermal evolution (in the orders of ~1 ns vs. ~1 ms, respectively) [55]. The validity of the two time-scale approximation is verified by the heuristic condition $\rho c_p d_{hs}^2 / k_{th} \sim \tau$, where d_{hs} is the hot spot width and $\tau = 2\pi / \omega$ is the wave period [36]. The EM bandwidth is also assumed to be sufficiently narrow to neglect the permittivity frequency dependence; hence, the wave equation (1) can be solved in the frequency domain while the heat equation (Equation 3.2) is computed in the slowly varying time domain. As the temperature rises, the spatial variations in the material properties ε and σ_c modify the microwave radiation pattern and hence enable the LMH self-focusing effect.

The thermal-runaway instability accelerates the local temperature growth rate (due to the temperature dependence of the material properties) in a way that the dielectric loss (i.e., the dissipated EM power) increases and/or the thermal conductivity decreases with the temperature growth. A positive dependence on the temperature of the dielectric loss factor $\varepsilon_r''(T)$ characterizes many nonconductive materials [64–68], as shown for instance in Figure 3.3 for soda-lime glass [69], Nylon™ (Ertalon-6®) [70] and mullite [71]. Likewise, in concrete, the temperature-dependent dielectric permittivity is numerically approximated by [72]

$$\varepsilon(T) = \varepsilon_0 \sum_{n=0}^{N}(a_n - jb_n)(T - T_{RT})^n, \tag{3.4}$$

where $T_{RT} = 300K$ and the best-fit polynomial coefficients for $N = 5$ are $a_{0-5} = 4.33$; $-6.9 \cdot 10^{-3}$; $2 \cdot 10^{-5}$; $-3 \cdot 10^{-8}$; 10^{-11}; $4 \cdot 10^{-15}$, and $b_{0-5} = 0.1$; $-0.7 \cdot 10^{-3}$; $4 \cdot 10^{-6}$; $-9 \cdot 10^{-9}$; $9 \cdot 10^{-12}$; $-2 \cdot 10^{-15}$. The other parameters are assumed constant, $k_{th} = 1.4$ W/mk, $c_p = 880$ J/kgK, and $\rho = 2,210$ kg/m^3 [73]. The simulated microwave frequency and the incident power are 2.45 GHz and 0.85 kW, respectively.

The microwave drilled region dissipates heat in various ways, including heat convection and radiation through the surface and heat conduction by the center electrode. The heat flux dissipated by convection is given by $q_{cond} = h_f (T - T_a)$, where h_f is the convective heat-transfer coefficient, and T_a is the ambient temperature. The blackbody radiation from the surface at high temperatures is assumed, $q_{rad} = \sigma_{sb} \varepsilon_s (T^4 - T_a^4)$, where σ_{sb} and ε_s are the Stephan-Boltzmann constant and the surface emissivity, respectively. Both the loss components of heat convection and radiation are subtracted from the power-density term in Equation 3.3 in the corresponding boundaries. The EM boundary conditions are chosen as either reflecting or absorbing according to the specific shielding structure employed.

Numerical models of the microwave-drill LMH effect were first derived as finite-difference time-domain (FDTD) simulations, and later by using commercial multi-physics software packages. The basic LMH model is mainly valid for the initial period of the hot spot formation, before melting, in the scheme depicted in Figure 3.1. In more advanced versions, the simulation calculates the temperature evolution in front of the microwave drill, including

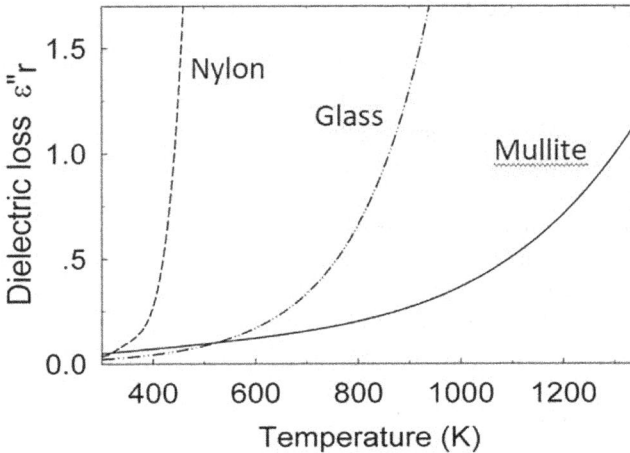

Figure 3.3 Temperature dependencies of the dielectric-loss factor approximated for Nylon™, glass, and mullite. The dielectric-loss increase with the temperature is essential for the LMH effect and for the feasibility of the microwave drill operation [36].

the latent heat; hence, it mimics the drill-bit insertion into the workpiece at the melting point, and updates the simulated geometry accordingly.

For the symmetrical microwave drill structure shown in Figure 3.1a, one may assume a cylindrical symmetry and a transverse-magnetic (TM) EM mode, which only consists of axial and radial electric-field components, E_z and E_r, respectively, and an azimuthal magnetic-field component $H\varphi$ (where z, r, and φ are the cylindrical coordinates shown in Figure 3.1a). The wave equation (Equation 3.1) is reduced to the Faraday and Ampere equations, as elaborated in Jerby et al. [55], while the heat equation (Equation 3.3) attains the form

$$\rho c_p \frac{\partial T}{\partial t} = k_{th}\left[\frac{1}{r}\frac{\partial}{\partial r}\left(r\frac{\partial T}{\partial r}\right)+\frac{\partial^2 T}{\partial z^2}\right]+\frac{dk_{th}}{dT}\left[\left(\frac{\partial T}{\partial r}\right)^2+\left(\frac{\partial T}{\partial z}\right)^2\right]+Q_d, \quad (3.5)$$

where the time-variable temperature profile, $T(t, z, r)$, is assumed axially symmetric. The second term in the right-hand side of Equation 3.5) represents the effect of the temperature-dependent heat conductivity, $k_{th}(T)$, which may further enhance the hot spot local confinement for $dk_{th}/dT < 0$.

Numerically, the two time-scale approximation allows a simplified solution of the EM-heat equations by two solvers. The EM-wave propagation is computed in a relatively short period of time, within each computation cycle of the thermal solver, while the material properties are assumed to be stationary during these short periods of EM computations. The resulting EM power-density absorbed, Q_d, provides the input for the thermal solver (of Equation 3.5). The temperature profile and the material parameters are updated in each computation cycle of the thermal solver. When the local temperature in front of the center electrode exceeds the melting point, the material properties and the microwave drill geometry are modified accordingly. The numerical simulation of the microwave drilling process can be implemented in a FDTD approach [74–76] as applied in Jerby et al. [55] and Grosglick et al. [62], and/or by the finite-element method (FEM) as in Meir and Jerby [36] (using a commercially available software).

Simulation results are presented in Figures 3.4a–c, for microwave drilling in concrete, using a coaxial applicator with 2- and 8-mm diameters of the electrode and outer cylinder, respectively [77]. The microwave frequency and the input power are 2.45 GHz and 0.85 kW, respectively. The center electrode is inserted 4 mm into the concrete in a fixed position. The thermal runaway occurs after approximately 18 s in this case, when the heating rate near the electrode tip is abruptly increased to the order of about 10^2K/s and higher, and the hot spot begins to relatively shrink. The hot spot localization is shown in Figures 3.4a–c at approximately 2 s after the thermal-runaway knee, by half-plane cross-section profiles of the electric-field distribution, the dissipated power density, and the temperature evolved. The electric field attains a maximal value of about 6 kV/cm. The absorbed microwave power results in only about 50 W total, due to the impedance mismatch and power

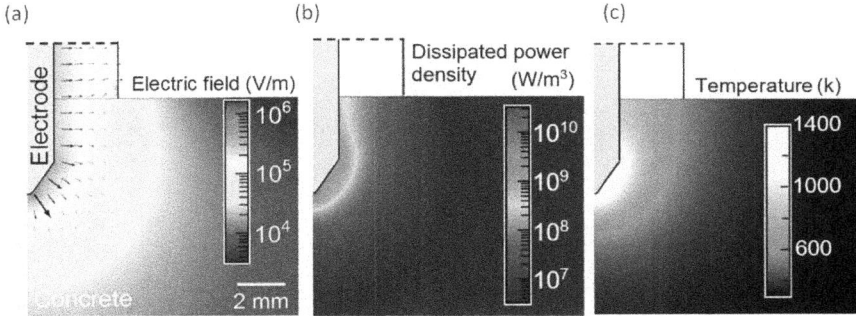

Figure 3.4 Simulated profiles (half-plane cross sections) of the electric-field localization (a), the dissipated power density (b), and the localized temperature (c), in the hot spot region around the center electrode inserted to concrete, after 20 s of 0.85-kW incident power at 2.45-GHz [77].

reflections (without tuning in this case). The temperature profile, exceeding the concrete's melting point, confirms the feasibility of a hot spot evolution in front of the electrode tip.

Another example shows simulation results of a 1-mm-thick soda-lime glass plate heated locally by 80 W at 2.1 GHz [36]. The glass softening temperature (~930 K) is reached near the electrode in 17 s. The electric field profile inside the glass exceeds 3.5×10^6 V/m in front of the electrode tip. The temperature distribution shows the hot spot confinement in a 1-mm° region, exceeding 930 K with the sharpened tip. The temperature profile evolution during the localized microwave heating results in an about 1–2-mm hot spot diameter. The thermal runaway is observed by a rapid rise of the peak temperature and hot spot formation accompanied by a drop in the initially high (~0.95) reflection coefficient.

The numerical simulation yields not only the temperature profile evolved in front of the microwave drill but also the time-varying input impedance and the reflection coefficient of the microwave drill toward the microwave source Jerby et al. [55], as given by

$$\Gamma_D(t) = \frac{Z_D(t) - Z_c}{Z_D(t) + Z_c}, \qquad (3.6)$$

where Z_c is the characteristic impedance of the coaxial waveguide and $Z_D(t)$ is the time-varying equivalent impedance of the microwave drill load, as illustrated in Figure 3.5. The equivalent load impedance consists of real and imaginary components (R_D and X_D, respectively) in the form

$$Z_D(t) = R_D + jX_D = R_{Heat} + R_{Rad} + j\left(\omega L - \frac{1}{\omega C}\right), \qquad (3.7)$$

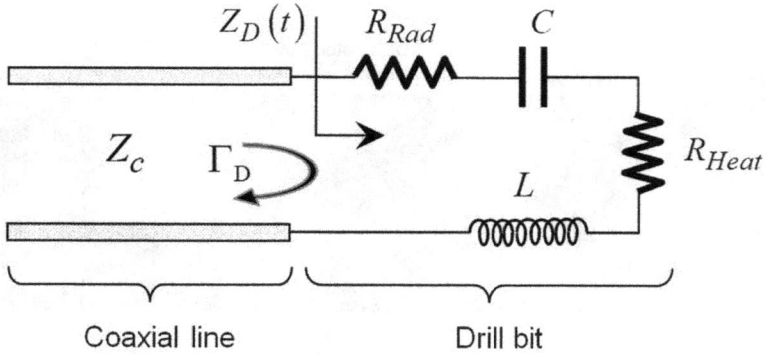

Figure 3.5 An equivalent circuit of the microwave drill input impedance. The interaction region within the drilled material is represented by equivalent resistive and reactive lumped elements, varying in time according to the spatiotemporal evolution of the microwave drilling process [55].

where R_{Heat} represents the effective power absorbed as heat in the hot spot region and R_{Rad} stands for the ineffective power component radiated outside the drilling spot. The reactive elements L and C represent the inductive and capacitive stored energies, respectively, in the microwave-drill near-field region (where the center-electrode insertion is limited to < $\lambda/4$ into the lossy material). During the microwave drilling process, $Z_D(t)$ may significantly vary from capacitive open circuit to resistive-inductive load, as the temperature increases and the electrode is inserted deeper into the hot spot.

In a more analytical approach, the equivalent components, R_{Heat}, R_{Rad}, L and C, can be approximately derived from an analytical model of a monopole antenna in a lossy uniform medium [78, 79] assuming a sine current distribution along the antenna. Using the Poynting theorem, the heating and radiated power components, as well as the stored (reactive) energy around the monopole antenna, can be found as described and analyzed in Jerby et al. [55]. In addition, a simplified heuristic condition for the hot spot initiation is proposed as $W>D$ [36, 80], where W is the net microwave power absorbed in the hot spot region, and D is given by

$$D = 2.5 k_{th} d_{hs} \Delta T_m / \Delta \varepsilon_r'', \tag{3.8}$$

where ΔT_m is the difference between the melting and room temperatures, and $\Delta \varepsilon_r''$ is the corresponding difference in the relative dielectric loss factor. The parameter D is roughly the net minimal microwave power needed to initiate the thermal-runaway process; hence, it is considered as a *drillability factor* in the context of the microwave drill operation. The condition $W > D$ can be satisfied in power levels of ~10–100 W for drilling a variety of practical materials, hence indicating the feasibility of low-power solid-state microwave drills, as presented in the next sections. For the sake of comparison, mullite, glass, and Nylon™ differ significantly in their properties, as reflected in their drillability factors, in the order of approximately 10 W, approximately 1 W, and approximately 0.1 W, respectively. Numerical simulations of microwave drilling at 0.1-kW input power into 1-mm thick slabs made of these materials yield time-to-melt (TTM) results of about 0.1 s in Nylon, about 10 s in glass, and about 60 s in mullite, according to their drillability factors [36].

Besides the physical insight, the various models provide computational tools for design and analysis of microwave drills, and for their real-time monitoring. For instance, the numerical analysis shows the critical need for an adaptive impedance matching [81], due to the rapid change in the load condition during its operation. The variation of the reflection coefficient Γ_D between the cold and molten hot spot extremes within seconds requires a model-based design of an adaptive impedance-matching tuner in order to continuously maximize the microwave-power transfer to the load (chasing the tuning condition $|\Gamma_D|^2 \to 0$) and to optimize the microwave drilling effectiveness during the entire operation. In the case of low-loss (*hard-start*) materials, such as sapphire or pure alumina for instance, the model may show the alleviating effect of a preheating stage, possibly by a local plasma generated by the same microwave drill applicator in a different mode [51].

3.4 Examples of Microwave Drilling in Various Materials

Microwave drills have been successfully applied to a variety of hard materials, including ceramics, concrete, basalt, glass, silicon, and plastics. This section presents several representative examples of experimental results that demonstrate the capabilities and limitations of the microwave drill technology in the aspects of the drilled materials, dimensions, qualities, and operating conditions.

3.4.1 Deep Holes in Concrete (Beyond ~λ/4 Depth)

Microwave applications have been developed for a variety of cement and concrete processing [82–84], including microwave-assisted curing processes [85], recycling of concrete demolition waste [86], concrete heating processes [87], and microwave furnaces for cement manufacturing. The basic microwave drill scheme presented in Section 2 is applicable for drilling and nail insertion (as in Figure 3.2) in concrete in the diameter range of approximately 0.5–5 mm and up to about 2.5 cm in depth. For larger hole diameters (in the order of ~1 cm and above) and for deeper holes, an advanced scheme has been developed, in which the outer cylinder of the coaxial structure is made as a hollow reamer that is also inserted into the hole and slowly rotated to deepen it while mechanically removing debris [77, 88]. Another approach for cutting concrete is based on the successive drilling of many small holes along the cut pattern [89].

Figure 3.6 shows the microwave drill extension that enables drilling holes much deeper than the ~λ/4 limit of the basic scheme presented in Figures 3.1a and 3.6a. In the more advanced version shown in Figures 3.6b through 3.6d, the outer coaxial cylinder also functions as a slowly-rotating hollow reamer, hence enabling the insertion of the entire coaxial applicator into the drilled substrate. In this mechanically assisted operation, the electrode tip is kept sufficiently close to the outer cylinder face, hence its optimal nearly λ/4 protrusion is also maintained in deeper holes. The debris is mechanically removed during the microwave drill operation in order to enable the entire drilling tube insertion, hence further deepening the hole. Referring to Figure 3.6c, the synchronized microwave drill operation includes the center electrode insertion (marked I) into the softened hot spot in order to maintain its optimal nearly λ/4 lead. The slow rotation of the outer cylinder (marked II) simultaneously cuts the softened concrete in the hot spot margin and further deepens the hole. The axial motion (marked III) inserts the entire coaxial structure into the extended hole. This motion is conditioned by the sufficient softening of the concrete in this region. Figure 3.6d shows the silver-coated hard-metal outer tube (12-mm outer diameter, 30-cm long), corrugated with helical grooves on its outside. With a movable center electrode inside, this coaxial structure transfers the microwave power to the drilled region and forms the hot spot. In addition, it is slowly rotated (at <20 RPM) hence

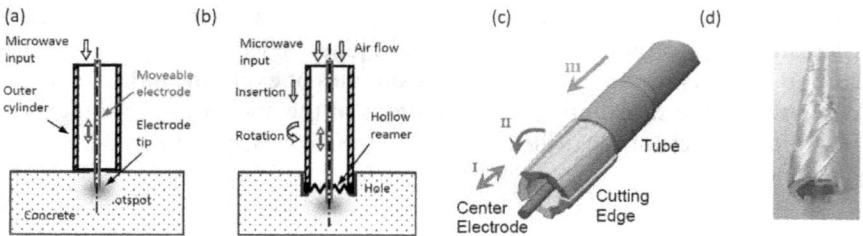

Figure 3.6 A microwave drill applicator for deep holes in concrete [77]: (a) A conceptual scheme of the basic microwave drill (Figure 3.1a) with a coaxial waveguide and a moveable center electrode inserted into the softened hot spot to form the hole; (b) an advanced microwave drill scheme for deeper holes, in which the hot spot is further extended to enable the removal of the softened debris by the slowly rotating outer cylinder; (c) the microwave drill tube used for deep holes in concrete as a hybrid of a coaxial open-end applicator and a hollow reamer. The center electrode is inserted into the hot spot (I) and further extend it. The debris are removed by the slowly rotating hollow reamer (II), enabling the axial progress of the entire drilling tube (III); (d) a realization of a silver-coated, 12-mmdiameter grooved outer cylinder.

it also functions as a hollow reamer removing the debris (as an option, pressurized air is blown through the coaxial structure in order to assist the debris removal and to cool down the tube).

A block diagram of a microwave-drill prototype for deep holes in concrete, depicted in Figure 3.7a, presents the main components of this microwave drill, including its microwaves, electronics and mechanical functions. The overview image in Figure 3.7b shows the 30-cm-long drill bit tube, the leading rail, the main unit, and the remote-control operating box. The 0.85-kW, 2.45-GHz magnetron is fed by a controllable switched-mode power supply (MagDrive 1000, Dipolar, Sweden, not shown here).

The microwave-drill system includes an embedded adaptive-impedance-matching unit, automatically operated with the other functions by a programmable controller embedded in the system. For example, the axial insertion motion (marked III in Figure 3.6c) is controlled by a feedback loop that measures the torque applied in the rotation motion (marked II there). This loop ensures the axial insertion below a certain torque threshold, as determined by the operator. This threshold also dictates the required level of the concrete softening, and consequently the level of noise emitted by the rotational reamer cutting (as a trade-off with the drilling speed). The remote handheld operating box (Figure 3.7b) maintains a serial communication with the main unit, displays the drilling progress and the actual operating conditions, and enables commands such as mode selection, power-level adjustment, and torque-limit setting.

The leading-rail basis is firmly attached to the drilled body and covered by a metallic foil to avoid microwave leakage. Microwave leak detectors are incorporated in the system in order to ensure the operator's safety. The

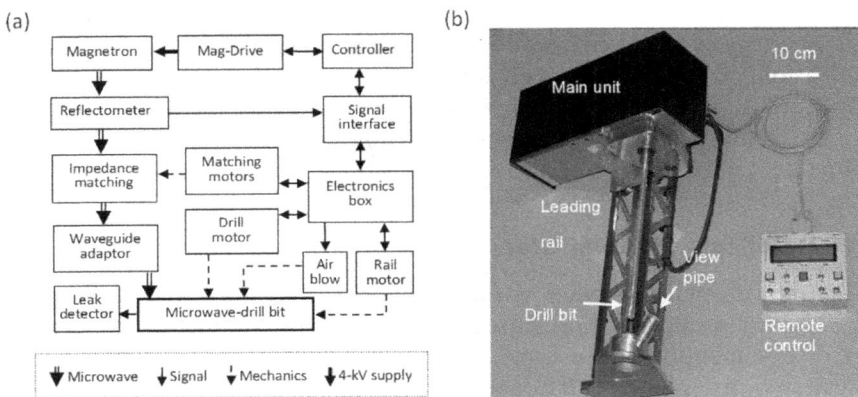

Figure 3.7 An overview of a portable microwave drill prototype for deep holes in concrete [77]: (a) A block diagram of the main microwave, electronics, and mechanical functions; (b) the experimental remotely operated prototype.

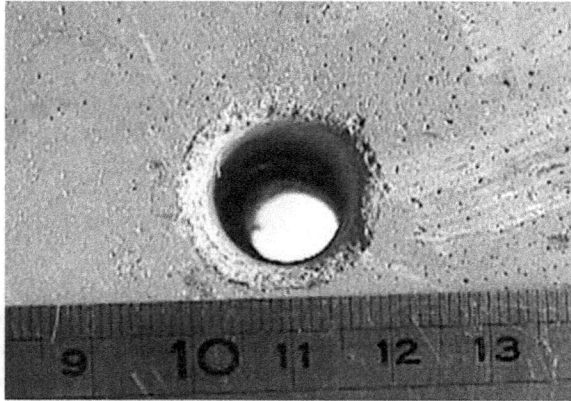

Figure 3.8 A 12-mm-diameter, 10-cm-deep hole in concrete made by a mechanically assisted microwave drill (Figure 3.7) capable of exceeding >26-cm-deep holes [77].

common safety standard for microwave emission from domestic and industrial installations (<1mW/cm^2([90] is satisfied in a proper microwave drill operation. The fixed positioning is also essential for a stable alignment in drilling deep holes (to reduce friction and avoid geometrical locking). The microwave drill makes no dust, vibrations or noise, and its operation is environmentally friendly.

This mechanically assisted microwave drill is capable of drilling 26-cm-deep, 12-mm-diameter holes in concrete, as shown in Figure 3.8. This record significantly extends the inherent nearly $\lambda/4$-depth capability of the basic microwave drill scheme. Compared to conventional mechanical drills, this microwave drill is characterized by a relatively silent and vibration-free operation, but its drilling speed is yet in the order of about 1 cm/min, significantly slower than mechanical drills. These performances can be useful for specific applications that critically require silent drilling operations in concrete.

3.4.2 Delicate Holes in Glass

The experimental feasibility of the LMH effect and microwave drilling of delicate holes in glass is presented here in the range of 10–100 W, using a solid-state microwave-generator apparatus [36]. The experimental setup is illustrated in Figure 3.9. It consists of a coaxial open-end applicator with a 1-mm° movable electrode, fed by a solid-state amplifier with positive feedback. The LDMOS-FET amplifier (Freescale MRF 6S21140) is tuned by a feedback loop to oscillate at about 2.1 GHz. It can generate up to

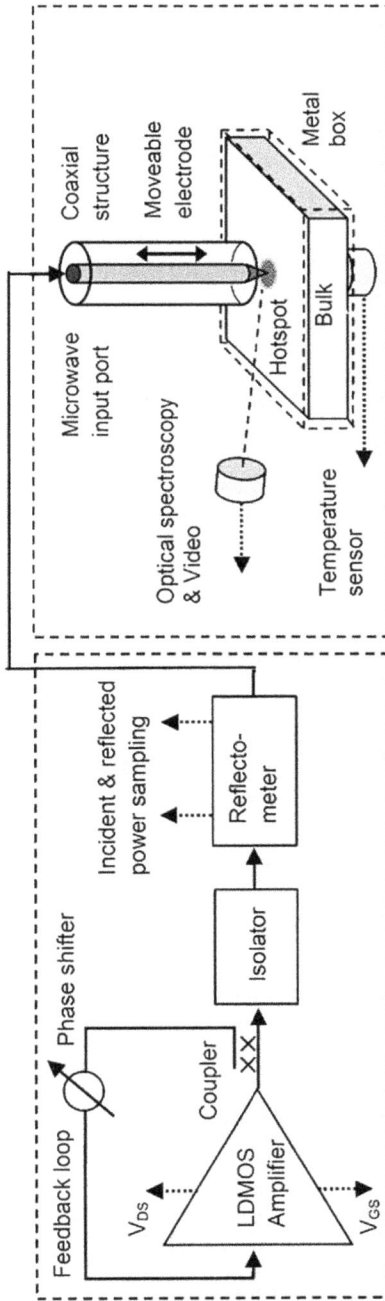

Figure 3.9 A transistor-based microwave drill experimental setup, fed by an LDMOS amplifier with a positive feedback loop [36].

140-W CW power, controlled by the V_{GS} and V_{DS} voltages. The transistor is protected from microwave reflections by an isolator. The incident and the reflected waves are detected by a reflectometer. The center electrode of the coaxial open-end applicator is penetrated into the softened hot spot evolved in the substrate material. An infrared temperature sensor (Raytek CI3) measures the temperature at the bottom surface of the heated material during the process. A video camera captures images of the local heating process via a side aperture.

Figure 3.10 shows measurements of the temperature evolution and the microwave reflection during the local heating of a 1-mm-thick soda-lime glass plate by 80-W incident power. The heating rate reveals the turning point of the induced thermal-runaway process ($t \sim 14$ s) accompanied by an abrupt decrease in the microwave reflection due to the better absorption of the microwave energy by the hotter region.

The molten hot spot can be observed in the transparent glass plate also via its rear surface. The temperature profile evolved is detected there by a thermal camera (FLIR E40) positioned as the temperature sensor in Figure 3.9. The thermal image and the temperature profile observed at the melting point are shown in Figure 3.11. The theoretically computed temperature profiles at the tip and rear planes show a good agreement with the measurements. The difference between the temperature profiles at the tip and the rear planes demonstrates the thermal diffusion through the 1-mm-thick glass plate. The hot spot confinement effect is clearly seen there.

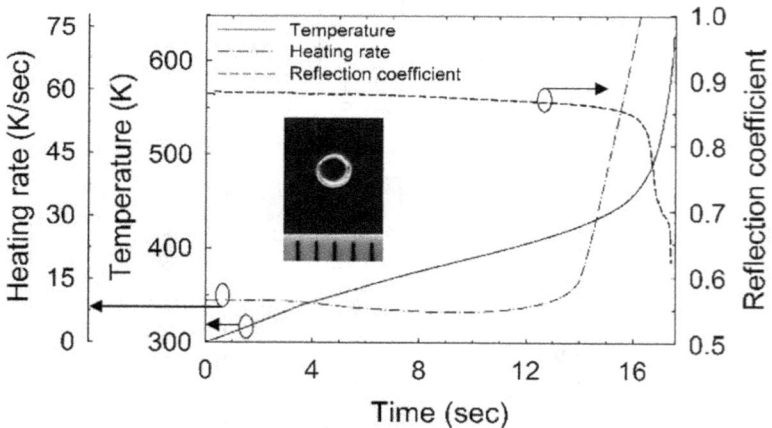

Figure 3.10 An experimental demonstration of the induced thermal-runaway instability in glass; the hot spot formation, melting and drilling, by 80-W of microwave power applied locally to a 1-mm glass plate. The temporal evolution of the temperature, the heating rate, and the microwave reflection coefficient show an abrupt change due to the instability at about 16 s. A typical about 1-mm$^\oslash$ hole made in these conditions is shown in the inset [36].

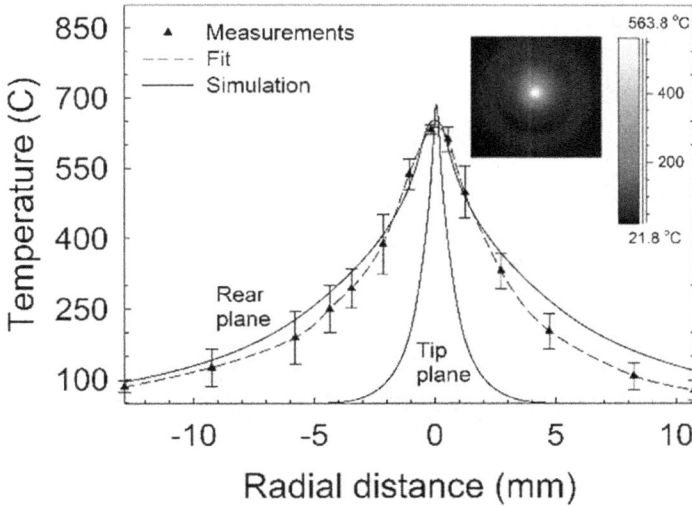

Figure 3.11 The temperature profile detected on the rear surface of the 1-mm thicktrans-
parent glass plate, by a thermal camera (FLIR E40), and its fit. The solid curves
show the numerical simulation result for the temparture profiles at the tip (z =
0.5 mm) and rear planes. The inset shows the FLIR image [36].

The microwave local heating effect, and in particular the drilling process,
can be enhanced by a slight plasma ejection from the heated surface toward
the electrode. After the initial stage of local microwave heating, the electrode
is lifted for a few seconds in order to generate a plasma column [36], which
assists the local pre-heating and alleviates the next stage of the microwave
drilling process. The additional plasma stage is shown in Figure 3.12 for
microwave drilling of a 4-mm thick glass plate by a graphite electrode at
80-W incident power. The various stages of the electrode manipulation and
plasma excitation are seen by the reflected microwave power.

A stepwise microwave drilling effect is observed in >2-mm deep inser-
tions. Figure 3.13 shows the effect in a 4-mm thick glass plate subjected to
70-W input power. The stepwise progress under a modest mechanical force
implies that two successive hot spots are formed during the process in two
different depths, as indicated by the microwave reflection drops. The first
hot spot enables the initial penetration, which could be continued further
only after the second hot spot has formed underneath. The stepwise microwave drilling, with two steps of ~2-mm each, provides also a rough measure
of the hot spot depth, which coincides with the theoretical results.

The microwave drill operation on brittle materials, such as silicon and
glass, may cause cracks due to thermal stresses. These were observed by

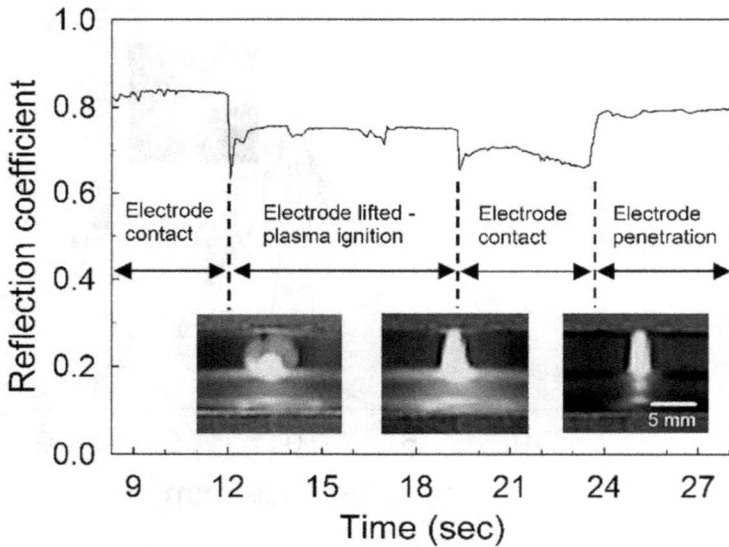

Figure 3.12 A microwave drilling process, enhanced by plasma ejection, in a soda-lime glass plate [36]. The various stages are seen by the reflection coefficient variations, and visually by the side-view images in the insets.

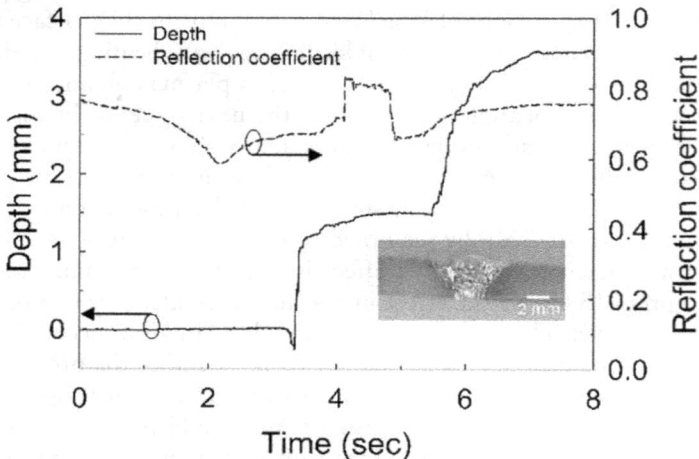

Figure 3.13 A stepwise microwave drilling in a 4-mm thick glass plate [36]. Each step is initiated by a hot spot that enables the electrode penetration to about a 2-mm depth. The progress is delayed then until another deeper hot spot is evolved and enables further penetration. The inset shows a crosscut in the glass plate.

optical and scanning electron microscopes in approximately 1–2-mm larger radii around some of the holes. These cracks are avoided by a more gradual operation and/or by preheating (e.g., by plasma). For larger holes, the confined cracked HAZ is removed mechanically during the microwave drilling process, as a means to intentionally enlarge the hole. Hence, the mechanically assisted microwave drilling method consists of repeating stages of applying the microwaves again on the same spot, cooling it down, and then mechanically removing the porous debris by a slight grinding. The removal of the soft, porous debris is performed, for instance, by a slowly rotating reamer, either separated or integrated (as the hollow reamer on the outer cylinder of the coaxial structure presented in Figure 3.6).

Figure 3.14 shows holes with approximate diameters of 1.6-mm and 2.4 mm and the arrays of four holes of with an approximately 0.5-mm diameter each, in about 1-mm-thick plates of borosilicate and soda-lime glass [26]. These results demonstrate the feasibility of making holes in these types of glass using the microwave drill with minimal radial cracking. Less attention is paid in these experiments to the hole shape accuracy (note that some imperfection is caused by the manual punching action). The approximately 0.5-mm-diameter holes represent the smallest diameter borings achieved to date by a 2.45-GHz (0.12-m wavelength) microwave drill.

Derivatives of microwave drill devices, designed for other applications involving glass and glass products, demonstrate the feasibility of glass cutting, tube sealing, and surface shaping [26, 36].

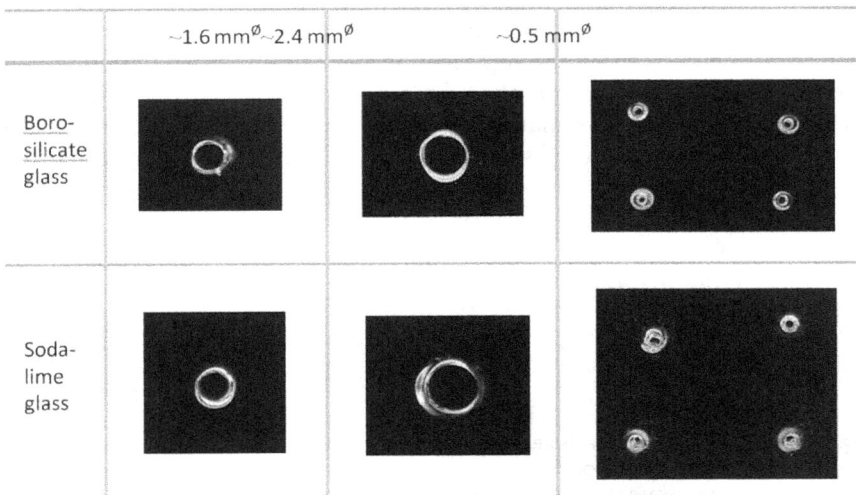

Figure 3.14 Microwave-drilled holes in 1-mm thick borosilicate and soda-lime glass plates [77].

3.4.3 Drilling and Pin Insertion in Ceramics

Microwave drills, as presented earlier, have also been tested on various ceramics, such a slow-purity alumina, nullite-cordierite, and glass ceramics [28, 29, 52]. Figure 3.15a shows for instance a 1-mm hole made in a plate of glass ceramic (Aremco, 502–600). The glossy debris were not removed in this example, in order to demonstrate the effect. The debris produced by the microwave drill in various materials are observed in various forms, that is, (a) compression of the solid material (if slightly porous) to the wall; (b) conversion to glass solidified inside or outside the hole, as in Figure 3.15; and (c) evaporation and gas emission. In ceramics and aggregated materials, the microwave drill may add a surficial sintering-like effect; thus, the wall could be strengthening there.

The microwave drill apparatus is also applicable for inserting and joining metallic or ceramic nails to ceramics. Figures 3.15b, c show examples of a 0.5-mm$^\varnothing$ tungsten nail inserted by the microwave drill into a plate of a zirconium phosphate (Aremco 502–1550), and an alumina tube inserted and joined by the microwave drill into a plate of aluminum silicate (Aremco 502–1100).

The microwave drill enables a distinction between different materials, and particularly between dielectrics and metals. For instance, as demonstrated for zirconia-based thermal-barrier coatings [29] with >1500°C melting temperature, the microwave drill can make holes and grooves in dielectric coatings with no damage to the underlying metallic substrate. The inherent material selectivity makes the microwave drill applicable for the controlled removal of ceramic coatings from underlying metallic substrates. For instance, the microwave drill process can make via holes through ceramic thermal barrier coatings (TBCs) to uncover an array of simulated cooling holes in advanced gas-turbine blades.

Figures 3.16a, b show the top-view of a 1.27-mm$^\varnothing$ hole, partially covered by the TBC, before and after the ceramic removal by the microwave drill (note the smooth circular exit hole created by the drill). Figures 3.17a and b show crosssections of smaller holes (1.02 mm$^\varnothing$),fully covered by the

Figure 3.15 (a) A 1-mm$^\varnothing$ via hole made by a microwave drill in a glass-ceramic plate, with the debris outflow (usually removed) remained on purpose in order to demonstrate the effect. (b, c) Pin insertion into ceramic plates: A tungsten nail (0.5mm$^\varnothing$) inserted into a zirconium-phosphate disk (b), and analumina tube (2.5-mm$^\varnothing$) inserted and joined to an aluminum-silicate disk (c) [28].

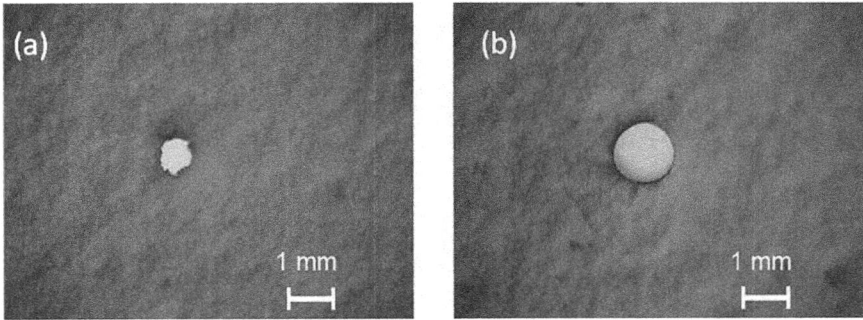

Figure 3.16 Top-views of a thermal-barrier coating (TBC) on a metal substrate, before (a) and after (b) the microwave drill exposure of an underlying 1.27-mm$^\varnothing$ hole in the metal substrate through a dense TBC [29].

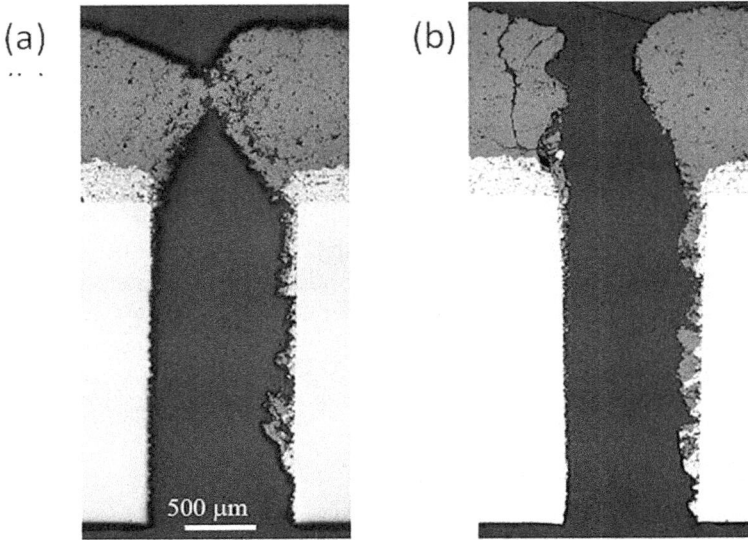

Figure 3.17 Cross-sections in the coated metal substrate (Hastelloy®X) with 1.02-mm$^\varnothing$ holes, as sprayed (a) and after their microwave drill exposure (b) [29].

TBC, and exposed by the microwave drill, respectively. The ceramic at the perimeter of the drilled hole is dense and cracked. This recrystallized yttria-stabilized zirconia (YSZ) indicates the lateral extent of the molten pool created during the drilling operation. The microwave operation is stopped after reaching the surface of the metal plate. Thus, the drill does not remove the TBC lying deep within the hole. In all holes examined, the microwave drill

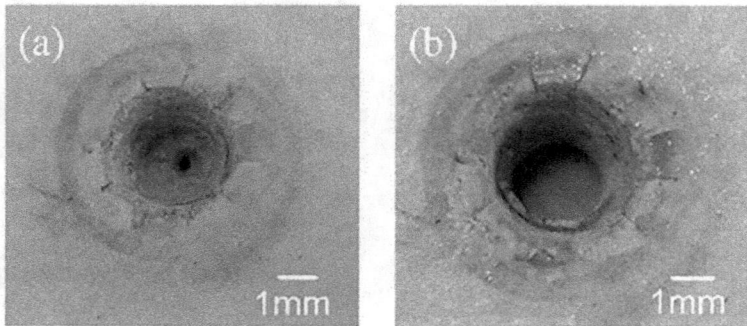

Figure 3.18 A stepwise mechanically-assisted microwave drilling of a 7-mm thick ceramic tile [36]. (a) The initial dimple after a slight mechanical removal of the porous region caused by the localized microwaves, and (b) the final hole obtained after four repetitive steps.

process does not seem to affect the microstructure of the underlying metallic substrate.

Another intentional stepwise mechanically assisted drilling procedure is demonstrated on a 7-mmthick, ceramic tile (Euro EN14411), in Figures 3.18a and 3.18b [28]. Such through-holes are drilled by applying approximately 0.1 kW incident power to a 1-mm$^{\varnothing}$ drill bit in four repeating steps. Each step includes the insertion of the drill bit into the substrate after the hot spot has formed (in a ~10-s TTM), and then a slight mechanical grinding to remove the debris. The stepwise drilling process helps avoid damage to the bulk, which could happen during continuous drilling due to thermal and mechanical stresses (though it does not prevent some imperfections as seen in Figure 3.18b).

3.4.4 Microwave Drilling In Bones

Drilling in bones for orthopedic purposes is a common practice, using a variety of drilling bits [9, 91]. Rotary drills are efficient, but they suffer several drawbacks, including debris and chips spread resulting in foreign-body reactions, substantial hematoma at the drilling site, heat generation, accuracy, and wobbling. The laser-based alternative for bone drilling [92] might be too costly for large-scale use in the clinical setting. The microwave drill has been studied therefore as a potential approach for drilling of bones [31].

In microwave interactions with biological soft tissues [45], at temperatures over about 50°C, the tissues undergo vaporization and carbonization. Higher temperatures may cause desiccation, protein denaturation, coagulation, and finally welding and cavitation. Microwave radiation may also have additional nonthermal effects on biological tissues, including a possible carcinogenicity [93]. Tumor ablation by RF radiation [94] is aimed to cause coagulation necrosis of cancerous tissue, thus tumor lysis and ablation. A percutaneous electrode (similar to a microwave drill) is inserted under imaging guidance (computed tomography, magnetic resonance imaging, or US) to ablate sub-dermal lesions. A similar method was also studied for ablation of tumors in bones [95].

An experimental study of microwave drilling in fresh wet bone tissue *in vitro* [31] showed a feasibility toward applicability studies for orthopedic surgery. The microwave drill employed approximately 0.2 kW incident power at 2.45 GHz, in order to penetrate bone at a drilling speed of about 1 mm/s (~5 s for a typical hole). The effective microwave power varied during the drilling process, as shown in Figure 3.19, due to the variations in the microwave load impedance, and in the consequent reflection from the bone, as the drilling process evolved.

The effect of microwave drilling on the mechanical properties of whole ovine tibial and chicken femoral bones drilled *in vitro* is seen by 3-point bending strength and fatigue tests, compared to similar mechanically drilled bones. The results show that microwave drilling does not induce a greater deteriorating effect on bone properties (including resistance to fatigue) than does mechanical drilling.

The carbonized margins around the microwave drilled hole occupy about 15% of the hole's diameter, as shown in Figure 3.20 [31]. SEM observations show that the microwave drill produces substantially smoother holes

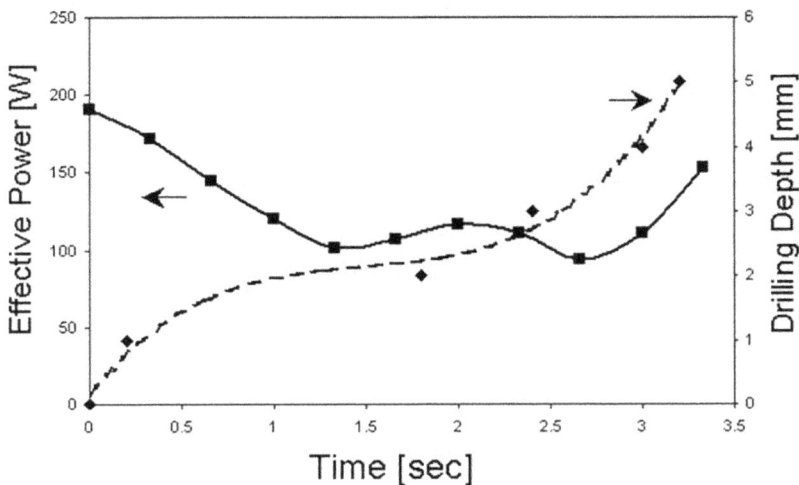

Figure 3.19 The microwave effective power (solid line) and drilling depth (dashed line) versus time during microwave drilling through a bovine trabecular bone [31].

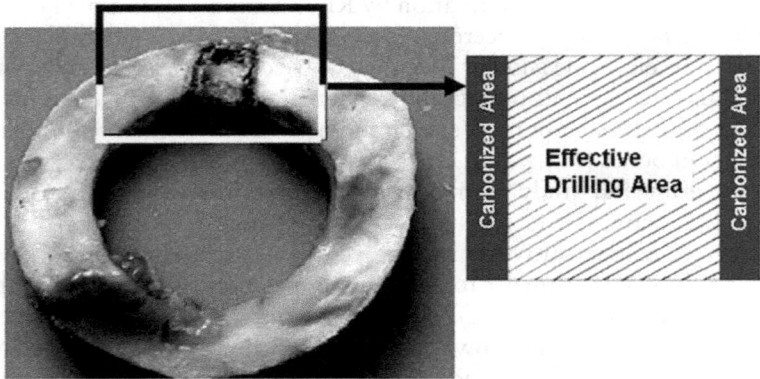

Figure 3.20 A transverse slice in the shaft of a cortical bone (ovine tibia with marrow exposed under the fractured bone), after microwave drilling *in vitro* [31]. The carbonized margins around the hole can be identified. The scheme on the right frame demonstrates the method of calculation of the carbonization effect ratio. The microwave drilling showed no visible thermal damage to the bone marrow.

(a) Microwave drill   (b) Mechanical drill

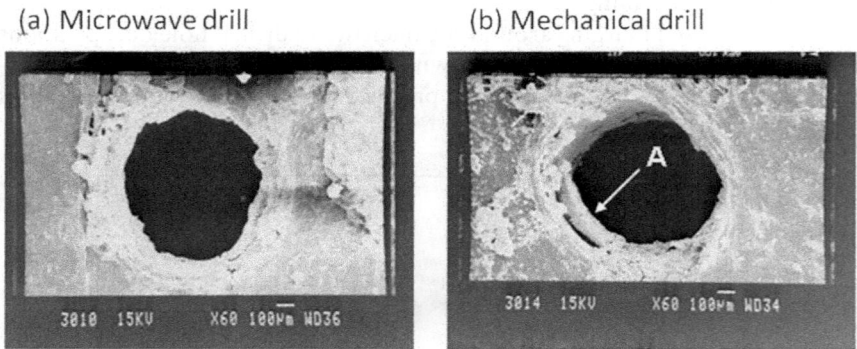

Figure 3.21 Scanning electron microscopy (SEM) images of chicken femora drilled with microwave (a) and mechanically (b) [31]. The hole geometry is substantially smoother in microwave drilling. Magnifications and scales are provided under each SEM frame.

in cortical bone than those produced by a mechanical drill, as shown in Figure 3.21 [31]. Typical defects in mechanical drilling are the large fragment of bone partially detached from the hole surface (marked A), sharp bone edges still attached to the contour of the hole, and scratches. Typical imperfections in microwave drilling are strut-like elements around the hole perimeter and small lumps, apparently formed by melted bone minerals. The hot spot produced by the microwave drill has the potential for overcoming two major problems presently associated with mechanical drilling in corti- cal and trabecular bones during orthopedic surgeries, namely, the formation of debris and the rupture of bone vasculature during drilling.

Safety of patients and staff is a critical issue to consider. A tailored design for the microwave drill and its screening for each specific surgical application are expected to significantly reduce the radiation exposure of both the patient and staff to a permitted minimum. For comparison, RF ablation procedures use similar power levels for about 10 minutes [94], much longer than needed for microwave drilling in bones. Nevertheless, research efforts are still required to determine whether microwave drilling induces a risk for carcinogenic effects and how such risk can be minimized in microwave drilling in bones.

3.5 Other Potential LMH Applications

The LMH intensification effect enables a wide range of applications in a vari- ety of fields [48]. Modified microwave drill applicators have been studied in a variety of LMH experiments, including e.g. additive manufacturing and 3D printing [57, 96, 97], joining, cutting [24, 56, 89], doping [38], dusty-plasma ejection [32, 51, 58], thermite ignition [59, 60], basalt melting [32], glass sealing [26], and microwave-induced breakdown spectroscopy (MIBS) [98]. Several applicative aspects are further discussed in more detail in this section.

3.5.1 LMH of Metal Powders

Coupling mechanisms of microwaves and metal powders are known in the literature in various volumetric schemes, reviewed in [57]. Recent experiments presented there show that metal powders with negligible dielectric losses can also be effectively heated and incrementally solidified by localized microwaves. This LMH effect is attributed to the magnetic component of the EM field, and to the eddy currents induced in the metal-powder particles. This effect is intensified by the micro-powder geometry, and it also occurs in diamagnetic metals such as copper. The heat is generated due to the metal electric resistivity, which impedes the eddy currents. In magnetic-like heating of metallic powders, the LMH effect is not characterized by thermal-runaway instability since the temperature tends to stabilize at about 700 K due to the particle necking and consolidation effects. The phenomenon of LMH confinement in metal pow- ders provides a basis for the applications presented in the next two sections.

3.5.2 Additive-Manufacturing and Three-Dimensional-Printing Processes

The LMH effect in metal powders is also associated with internal micro- plasma breakdowns between the particles, which leads to local melting and solidification of the metal powder. This effect enables an LMH-based

technique for stepwise three-dimensional (3D) printing and additive manufacturing (AM) of metal parts from powder [57, 96, 97]. The consolidated
drop of metal powder is used in this technique as a building block, placed on
top of the previously constructed structure, in a stepwise AM process. The
present results show rods constructed by LMH-AM from bronze-based and
iron powders. The feasibility of magnetic fixation (instead of mechanical
supports) is demonstrated for iron powder [96].

3.5.3 Ignition of Thermite in Air and Underwater

Powder mixtures, such as of pure aluminum and magnetite (or hematite),
may generate energetic thermite reactions. These could be useful for a variety
of combustion and material processing applications. However, their usage
is yet limited by the difficult ignition of these reactions. Ignition of thermite
reactions is found feasible by intensified LMH in air [59] and underwater [60]. The power required for thermite ignition by LMH is about 0.1
kW for a approximately 3-s period, provided by a solid-state microwave
generator. The thermite mixture exhibits both dielectric and magnetic loss
mechanisms. The magnetic LMH is implemented by a short-end applicator,
which enhances the magnetic field in front of it. It yields a faster heating rate
than the open-end, dielectric-LMH applicator, up to the Curie temperature
at 858K, where the magnetic losses significantly decrease. Integrating both
magnetic and electric LMH mechanisms by a hybrid applicator enables the
thermal-runaway instability and the thermite ignition. These experiments
also demonstrate the feasibility of cutting and welding by relatively low-
power LMH.

The initiation of the intense exothermic reaction in thermites, also demonstrates an example of LMH ignition of other high-temperature self-
propagating syntheses (SHS). Due to their zero-oxygen balance, exothermic
thermite reactions also occur underwater. However, this feature is also difficult to utilize because of the hydrophobic properties of the thermite powder
and its tendency to agglomerate on the water surface rather than to sink
into the water. The recently discovered bubble-marble (BM) effect enables
the insertion and confinement of a thermite-powder batch into water by a
static magnetic field, and its ignition by LMH underwater [60]. Potential
applications of this underwater combustion effect may include wet welding, thermal drilling, detonation, thrust generation, material processing,
and composite-material production. These could be implemented in other
oxygen-free environments as well, such as the outer space.

3.5.4 Dusty-Plasma Ejection from Solids

Dusty plasmas in forms of fireballs and fire columns can be ejected by LMH
directly from hot spots evolved in solid substrates, as presented, for example,

in Jerby and colleagues [32, 51, 58] for various dielectric and metallic materials. The intensified LMH-plasma process begins with a hot spot formation as presented earlier (e.g., for microwave drilling). However, for plasma ejection the electrode is lifted up (rather than pushed in) in order to detach the molten drop from the surface, and to further inflate it to a form of a buoyant fireball.

Beside their resemblance to natural ball-lightning phenomena, fireballs and fire-columns, may also have practical importance, for example, as means to produce nanoparticles directly from various substrate materials, such as silicon, glass, ceramics, copper, titanium, and others [32, 51, 58]. Nanoparticles were observed in these and other materials, both by *in situ* synchrotron small-angle X-ray scattering (SAXS) of the dusty plasma and by *ex situ* SEM observations of the nano-powders collected after the processes.

3.5.5 Material Identification by Microwave-Induced Breakdown Spectroscopy

The LMH-generated plasma-column, ejected from a hot spot in the material (to be identified), can be used for an atomic emission spectroscopy of the light emitted by the plasma. The proposed microwave-induced breakdown spectroscopy (MIBS) [98] is similar to the laser-induced (LIBS) identification technique, except that the plasma is excited by LMH rather than by laser.

3.5.6 Surface Treatment and Doping

The microwave drill concept, applied for localized heating of silicon plates in a heating rate of >200 K/s up to the melting point [37], also enables localized thermal processes in silicon, including joining, welding, drilling, and doping. The latter is demonstrated by local doping of silicon substrates by LMH, using silver and aluminum dopants [38]. The dopant material is incorporated in these processes in the electrode tip and is diffused into the locally heated bulk to form a submicron PN junction.

Chemical reactions applied by LMH intensification for surface treatments also include thermite reactions for the conversion of rust to iron and alumina [59]. These shallow LMH techniques open new possibilities for various local surfical processes.

3.5.7 Cutting

While cutting in glass, ceramics, concrete, and other materials by a modified microwave drill apparatus could be conceived as a straightforward extension of the drilling operation [24, 89], cutting in metal bulks by LMH is more challenging due to the metal's high conductivity. A recent experimental study [56] shows that adding a relatively small direct current (DC) may catalyze an LMH effect in the iron bulk, up to its local melting (and further to ablation and dusty-plasma ejection). The combined DC-LMH effect is demonstrated there by cutting 8-mm$^\varnothing$ iron rebars (made of carbon steel) with no susceptors added. The synergic microwave-DC effect is attributed to a combined thermal skin evolution, which jointly forms a hot spot in a mutually intensified thermal-runaway instability. A simplified theoretical model of the combined DC-LMH interaction in iron demonstrates the evolution of the thermal skin layer and confirms the experimental observations. This synergic DC-LMH effect may advance microwave applications in thermal processing of metals (e.g., melting, cutting, joining, sintering, casting, and 3D printing).

3.5.8 Heat Intensification in a Microwave Cavity

LMH intensification due to the temperature-dependent material properties is demonstrated by irradiating basalts in a microwave cavity [30, 32]. The heat intensification effect is clearly seen as the basalt brick is melted inside, while its outer surface remains solid. The molten core ejects lava outside the brick, which leaves a void inside (whereas the outer surface of the brick remains solid). A 3D multiphysics model, using Equations 3.1 through 3.3), agrees well with the temperature profiles measured on the basalt brick faces. Similar demonstrations are also conducted with natural-shape basalt stones to mimic various volcanic phenomena in a laboratory scale. This LMH effect can be further used in order to intensify processes of mineral extraction from rocks.

3.6 Conclusion

The various LMH aspects reviewed in this chapter differ from the common microwave heating paradigm by the intentional concentration and local intensification of microwave heating energy into a HAZ, much smaller in size than the microwave wavelength. The relatively high power density concentrated within this small volume causes the hot spot melting, evaporation, and even breakdown to plasma. The LMH concept differs from the uniform microwave heating paradigm, by its physical mechanism, theoretical considerations, technical implementations, and the needs that it serves. As such, the LMH forms a new paradigm [47, 48] that expands the field of microwave heating to relatively small but much more intensified HAZs.

The theoretical, coupled-thermal-electromagnetic model of the subwavelength LMH interaction confirms the hot spot evolution in front of its applicator due to the intentional thermal-runaway instability. The role of the temperature-dependent material properties on the confined heating process is identified, and the consequent variation in the microwave load impedance is quantified (due to the spatiotemporal temperature evolution and the varying geometry). The theoretical analysis provides both insight to the physical principles of the LMH mechanism and a valid model for LMH applicator designs in various materials and their operational monitoring.

The microwave drilling concept described here, as representing an application of the broader LMH paradigm, has been successfully demonstrated on a variety of materials, for example, concrete, glass, silicon, basalt, various ceramics, and bones. The microwave drill operation is silent and gentle, and it produces no vibration. However, its speed, accuracy, and versatility are yet inferior with respect to conventional drilling techniques, and it requires more research and development efforts in these aspects. An increase in the microwave drilling speed, for instance, could be achieved by one or more of the following means: (a) non-LMH preheating, for example, by a local plasma; (b) a more efficient LMH applicator; (c) a faster adaptive impedance-matching tuner; and/or (d) a quicker debris-removal process.

The silent operation makes the microwave drill relevant to construction works in sensitive areas where loud noise and vibrations must be avoided (e.g., residential and office areas, schools, hospitals, etc.). The microwave drill technology also utilizes available components of domestic microwave ovens and cellular systems; hence, it may provide affordable solutions for construction and maintenance work. It may also be integrated in more specialized industrial applications; in drilling, cutting, and jointing operations; and additive manufacturing processes. LMH may also assist mining operations by inducing thermal stresses and local weakening of rocks [15, 24, 30, 32].

Safety concerns and considerations of radio-frequency interference (RFI) impose limitations on the microwave drill operation. The shielding required to comply with safety standards could be performed either by a closed chamber (as the domestic oven), or as a soft shielding cover. The closed structure is more applicable for automatic production lines, whereas the flexible shield may fit in construction and geological works. Both solutions, experimented in practice, satisfy <1 mW/cm^2 microwave power leakage, in accord with common safety standards [90].

The microwave drill is also found applicable for various ceramic materials, such as low-purity alumina, zirconia, mullite, and glass ceramics. The selectivity feature of the microwave drill was found useful for punching cooling holes in ceramic TBC (e.g., as used to protect jet-engine blades) in order to expose the underlying cooling holes without damaging the metallic substrate [29].

The theoretical and experimental studies earlier demonstrate the applicability of relatively low-power microwaves, in the <0.1-kW range, for localized heating and melting of materials like concrete, glass, ceramics and basalts. Due to the emerging cellular technologies, adequate microwave sources in this range are available by various solid-state devices [99], such as gallium nitride (GaN) [100], silicon carbide (SiC) [101], and silicon-based

LDMOS [102] transistors. Accordingly, several microwave drilling experiments [35, 36, 89] demonstrate the feasibility of advanced solid-state microwave drills as compact, integrated and controllable devices.

The effectiveness of solid-state microwave drills could be significantly improved by an integrated adaptive impedance-matching tuner, in order to cope with the dynamically varying load from high to low impedance within seconds. An adaptive impedance-matching tuner, as common in magnetron systems [81], may either reduce the total power required to reach the same performance, or accelerate the heating process with the given power. Further improvement of the localized heating process could be achieved by a localized plasma preheating stage. This may extend the microwave drill abilities further to hard-start materials, such as pure alumina and various metals.

In future microwave drill systems, the solid-state implementation will enable real-time analyses of the reflected waves, scattered by the evolved hot spot load, and hence *sense* the underlying material and its current status. Gershon et al. [103] presents, for instance, a method to measure the actual dielectric properties (and temperature) by an open-end structure. This method, integrated with a numerical estimator based on the theoretical model presented earlier, could be used to monitor the microwave drilling status. It could also be extended to distinguish between different materials if expected in the drilled body (e.g., metal rebars in concrete walls), and to guide the drill bit accordingly. Hence, advanced microwave drills may have an additional *radar* feature, which may also detect the drilling conditions in complex structures. The microwave drill can be operated not only as a stand-alone tool but also in integration with other machining instruments. This may lead to the development of a *microwave-assisted-machining* (MAM) paradigm, in which microwaves would be incorporated in other machining operations.

The LMH intensification technique can be considered, to some limited extent, as a low-end substitute for laser-based applications, such as drilling and cutting, joining, surface treatment, material identification, and AM. While LMH may provide low-cost, compact, and efficient solutions in this regard, it requires a physical contact with the object (unlike the remote laser), and its resolution (~1 mm) is yet inferior with respect to lasers. Therefore, one may deduce that LMH applications, such as the microwave drill, are more relevant to operating regimes of relatively large volumes and rough processes, or as complementary means to the more accurate and expensive laser-based systems.

Acknowledgments

This research was supported in part by the Israel Science Foundation (Grant Nos. 1270/04, 1639/11, 1896/16 and 2504/19) and by the Israeli Ministry of Science and Technology, MOST (Grant No. 3–15631).

References

[1] K. Krajick, "New drills augur a great leap downward," *Science*, Vol. 283, no. 5403, pp. 781–783, 1999. https://doi.org/10.1126/science.283.5403.781

[2] W. Schulz, U. Eppelt, and R. Poprawe, "Review on laser drilling I: Fundamentals, modeling, and simulation," *Journal of Laser Applications*, Vol. 25, no. 1, p. 012006, 2013. https://doi.org/10.2351/1.4773837

[3] Y. Bar-Cohen, X. Bao, M. Badescu, S. Sherrit, K. Zacny, N. Kumar, T. Shrout, and S. Zhang, "High-temperature drilling mechanisms," *High-Temperature Materials and Mechanics*, Bar-Cohen, Y. (Ed.), CRC-Press, Boca Raton FL, 2014.

[4] J. Kumar, "Ultrasonic machining: A comprehensive review," *Machining Science and Technology*, Vol. 17, no. 3, pp. 325–379, 2013. https://doi.org/10.10 80/10910344.2013.806093

[5] V. Peržel, S. Hloch, H. Tozan, M. Yagimli, and P. Hreha, "Comparative analysis of abrasive waterjet (AWJ) technology with selected unconventional manufacturing processes," *Int'l Journal of Physical Sciences*, Vol. 6, no. 24, pp. 5587–5593, 2011, DOI:10.5897/IJPS11.460

[6] Y. Bar-Cohen, and K. Zacny, *Drilling in Extreme Environments: Penetration and Sampling on earth and Other Planets*, Wiley-VCHVerlag GmbH, Weinheim.

[7] W. C. Maurer, *Novel Drilling Techniques*, Pergamon Press, New York, 1979.

[8] W. C. Maurer, *Advanced Drilling Techniques*, Petroleum Publishing, Tulsa, 1980.

[9] K. L. Wiggins, and S. Malkin, "Drilling of bone," *Journal of Biomechanics*, Vol. 9, no. 9, pp. 553–559, 1976. https://doi.org/10.1016/0021-9290(76)90095-6

[10] E. Jerby, V. Dikhtyar, O. Aktushev, and U. Grosglick, "The microwave drill," *Science*, Vol. 298, no. 5593, pp. 587–589, 2002. https://doi.org/10.1126/science.1077062

[11] J. Thuéry, *Microwaves: Industrial, Scientific, and Medical Applications*, Artech House, Boston, MA, 1992.

[12] A. C. Metaxas, *Foundations of Electroheat: A Unified Approach*, John Wiley & Sons, Chichester, GB, 2000.

[13] S. Chandrasekaran, S. Ramanathan, and T. Basak, "Microwave material processing: A review," *AIChE Journal*, Vol. 58, no. 2, pp. 330–363, 2011. https://doi.org/10.1002/aic.12766

[14] J. Kim, S. C. Mun, H.-U. Ko, K.-B. Kim, M. A. Khondoker, and L. Zhai, "Review of microwave assisted manufacturing technologies," *International Journal of Precision Engineering and Manufacturing*, Vol. 13, no. 12, pp. 2263–2272, 2012. https://doi.org/10.1007/s12541-012-0301-2

[15] D. P. Lindroth, R. J. Morrell, and J. R. Blair, *Microwave assisted hard rock cutting*. United States Patent 5,003,144, 1991.

[16] J. F. Ready, *Industrial Applications of Lasers*. Academic Press, New York, NY, 1997.

[17] G. Roussy, A. Bennani, and J. M. Thiebaut, "Temperature runaway of microwave irradiated materials," *Journal of Applied Physics*, Vol. 62, no. 4, pp. 1167–1170, 1987. https://doi.org/10.1063/1.339666

[18] G. A. Kriegsmann, "Thermal runaway in microwave heated ceramics: A one-dimensional model," *Journal of Applied Physics*, Vol. 71, no. 4, pp. 1960–1966, 1992. https://doi.org/10.1063/1.351191

[19] P. E. Parris, and V. M. Kenkre, "Thermal runaway in ceramics arising from the temperature dependence of the thermal conductivity," *Physica Status Solidi (b)*, Vol. 200, no. 1, pp. 39–47, 1997. https://doi.org/10.1002/1521-3951 (199703)200:1<39::aid-pssb39>3.0.co;2-r

[20] G. Kriegsmann, "Hot spot formation in microwave heated ceramic fibres," *IMA Journal of Applied Mathematics*, Vol. 59, no. 2, pp. 123–148, 1997. https://doi.org/10.1093/imamat/59.2.123

[21] C. A. Vriezinga, "Thermal runaway in microwave heated isothermal slabs, cylinders, and spheres," *Journal of Applied Physics*, Vol. 83, no. 1, pp. 438–442, 1998. https://doi.org/10.1063/1.366657

[22] C. A. Vriezinga, S. Sánchez-Pedreño, and J. Grasman, "Thermal runaway in microwave heating: A mathematical analysis," *Applied Mathematical Modelling*, Vol. 26, no. 11, pp. 1029–1038, 2002. https://doi.org/10.1016/s0307-904x(02)00058-6

[23] X. Wu, J. R. Thomas, and W. A. Davis, "Control of thermal runaway in microwave resonant cavities," *Journal of Applied Physics*, Vol. 92, no. 6, pp. 3374–3380, 2002. https://doi.org/10.1063/1.1501744

[24] E. Jerby, and V. Dikhtyar, *Method and device for drilling, cutting, nailing and joining solid non-conductive materials using microwave radiation*. United States Patent US 6,114,676, 2000.

[25] E. Jerby, and V. Dikhtyar, "Drilling into hard non-conductive materials by localized microwave radiation," *Advances in Microwave and Radio Frequency Processing*, 687–694, Willert-Porada M. (Ed.), Springer, Berlin, 2006. (Proc. AMPERE-8 Int'l Conf. Microwave & RF Heating, Bayreuth, Sept. 4–7, 2001). https://doi.org/10.1007/978-3-540-32944-2_75

[26] E. Jerby, O. Aktushev, V. Dikhtyar, P. Livshits, A. Anaton, T. Yacoby, A. Flax, A. Inberg, and D. Armoni, "Microwave drill applications for concrete, glass and silicon," Microwave and Radio-Frequency Applications 4th World Congress, Conf. Proc., 156–165, Folz, D. C., & Schulz, R. L. (Eds). Austin, Texas, Nov. 4–7, 2004.

[27] E. Jerby, and V. Dikhtyar, "The microwave drill: A new technology will enable silent drilling and nailing in concrete, ceramics, and rocks," *Water-Well Jour.*, pp. 32–36, July 2003.

[28] E. Jerby, V. Dikhtyar, and O. Aktushev, "Microwave drill for ceramics," *Amer. Ceramic Soc. Bulletin*, Vol. 82, pp. 35–37, 2003.

[29] E. Jerby, and A. M. Thompson, "Microwave drilling of ceramic thermal barrier coatings," *Jour. Amer. Ceram. Soc.*, Vol. 87, pp. 308–310, 2004.

[30] E. Jerby, V. Dikhtyar, and M. Einat, "Microwave melting and drilling of basalts," AIChE Annual Meeting, Austin, Texas, Nov. 4–7, 2004.

[31] Y. Eshet, R. Mann, A. Anaton, T. Yacoby, A. Gefen, and E. Jerby, "Microwave drilling of bones," *IEEE Trans. Biomedical Eng.*, Vol. 53, pp. 1174–1182, 2006.

[32] E. Jerby, and Y. Shoshani, "Localized microwave-heating (LMH) of basalt: Lava, dusty-plasma, and ball-lightning ejection by a 'miniature volcano'," Scientific Reports, Art. No. 12954, 2019.

[33] N. K. Lautre, A. K. Sharma, and P. Kumar, "Distortions in hole and tool during microwave drilling of perspex in a customized applicator," *2014 IEEE MTT-S International Microwave Symposium (IMS2014)*, 2014, pp. 1–3, doi: 10.1109/MWSYM.2014.6848410.

[34] A. Singh, and A. K. Sharma, "On microwave drilling of metal-based materials at 2.45 GHz," *Appl. Phys. A*, Vol. 126, no. 822, 2020. https://doi.org/10.1007/s00339-020-03994-5

[35] O. Mela, and E. Jerby, "Miniature transistor-based microwave drill," *Proc. Global Congress Microwave Energy Applications*, Otsu, Japan, pp. 443–446, 2008.

[36] Y. Meir, and E. Jerby, "Localized rapid heating by low-power solid-state microwave-drill," *IEEE Trans. Microw. Theory Tech.*, Vol. 60, pp. 2665–2672, 2012.

[37] R. Herskowits, P. Livshits, S. Stepanov, O. Aktushev, S. Ruschin, and E. Jerby, "Silicon heating by a microwave-drill applicator with optical interferometric thermometry," *Semiconductor Science and Tech.*, Vol. 22, pp. 863–869, 2007.

[38] P. Livshits, V. Dikhtyar, A. Inberg, A. Shahadi, and E. Jerby, "Local doping of silicon by a point-contact microwave applicator," *Microelectron. Eng.*, Vol. 88, pp. 2831–2836, 2011.

[39] F. Marken, Y. C. Tsai, B. A. Coles, S. L. Matthews, and R. G. Compton, "Microwave activation of electrochemical processes: Convection, thermal gradients and hot spot formation at the electrode solution interface," *New J. Chem.*, Vol. 24, pp. 653–658, 2000.

[40] L. J. Cutress, F. Marken, and R. G. Compton, "Microwave-assisted electroanalysis: A review," *Electroanalysis*, Vol. 21, pp. 113–123, 2009.

[41] I. Longo, and A. S. Ricci, "Chemical activation using an open-end coaxial applicator," *J. Microw. Power Electromagn. Energy*, Vol. 41, pp. 1–4, 2007.

[42] G. B. Gentili, M. Linari, I. Longo, and A. S. Ricci, "A coaxial microwave applicator for direct heating of liquids filling chemical reactors," *IEEE Trans. Microwave Theory Tech.*, Vol. 57, pp. 2268–2275, 2009.

[43] T. Z. Wong, and B. S. Trembly, "A theoretical model for input impedance of interstitial microwave antennas with choke," *Int. J. Radiat. Oncol. Biol. Phys.*, Vol. 28, pp. 673–682, 1994.

[44] I. Longo, G. B. Gentili, M. Cerretelli, and N. Tosoratti, "A coaxial antenna with miniaturized choke for minimally invasive interstitial heating," *IEEE Trans. Biomed. Eng.*, Vol. 50, pp. 82–88, 2003.

[45] A. Copty, F. Sakran, M. Golosovsky, and D. Davidov, "Low-power near-field microwave applicator for localized heating of soft matter," *Appl. Phys. Lett.*, Vol. 24, pp. 5109–5111, 2004.

[46] C. L. Brace, "Microwave ablation technology: What every user should know," *Curr. Probl. Diagn. Radiol.*, Vol. 38, pp. 61–67, 2009.

[47] Y. Meir, and E. Jerby, "The localized microwave-heating (LMH) paradigm: Theory, experiments, and applications," Proc. GCMEA-2, July 23–27, 2012, Long Beach, California, pp. 131–145.

[48] E. Jerby, "Localized microwave-heating intensification: A 1-D model and potential applications," *Chemical Engineering and Processes (CEP)*, Vol. 122, pp. 331–338, 2017.

[49] N. Kondo, H. Hyuga, H. Kita, and K. Hirao, "Joining of silicon nitride by microwave local heating," *Jour. Ceram. Soc. Jap.*, Vol. 118, pp. 959–962, 2010.

[50] S. R. Wylie, A. I. Al-Shamma'a, and A. Shaw, "A microwave plasma drill," IET Conf. High Power RF Tech., London, pp. 1–3, 2009.

[51] E. Jerby, A. Golts, Y. Shamir, S. Wonde, J. B. A. Mitchell, J. L. LeGarrec, T. Narayanan, M. Sztucki, D. Ashkenazi, and Z. Barkay, "Nanoparticle plasma ejected directly from solid copper by localized microwaves," *Appl. Phys. Lett.*, Vol. 95, 191501, 2009.

[52] X. Wang, W. Liu, H. Zhang, S. Liu, and Z. Gan, "Application of microwave drilling to electronic ceramics machining," 7th Int'l Conf. Electronics Packaging Technology, Proc. Article # 4198945 (4 pages), Aug. 26–29, 2006, Shanghai, China.

[53] W. A. G. Voss, "Solid-state microwave oven development," *Jour. Microwave Power*, Vol. 21, pp 188–189, 1986.

[54] E. Schwartz, A. Anaton, D. Huppert, and E. Jerby, "Transistor-based miniature microwave heater," Proc. IMPI 40th Annual Int'l Microwave Symp., Boston, pp. 246–249, 2006.

[55] E. Jerby, O. Aktushev, and V. Dikhtyar, "Theoretical analysis of the microwave-drill near-field localized heating effect," *Jour. Appl. Phys.*, Vol. 97, 034909, 2005.

[56] Y. Shoshani, and E. Jerby, "Local melting and cutting of iron bulks by a synergic microwave-DC thermal skin effect," *Applied Physics Letters*, Vol. 118, Art. No. 194102, 2021.

[57] E. Jerby, Y. Meir, A. Salzberg, E. Aharoni, A. Levy, J. Planta Torralba, and B. Cavallini, "Incremental metal-powder solidification by localized microwave-heating and its potential for additive manufacturing," *Additive Manufacturing*, Vol. 6, pp. 53–66, 2015.

[58] E. Jerby, "Microwave-generated fireballs," Encyclopaedia of Plasma Technology, Taylor and Francis, 2017, pp. 819–832.

[59] Y. Meir, and E. Jerby, "Thermite-powder ignition by electrically-coupled localized microwaves," *Combustion and Flame*, Vol. 159, pp. 2474–2479, 2012.

[60] Y. Meir, and E. Jerby, "Underwater microwave ignition of hydrophobic thermite powder enabled by the bubble-marble effect," *Applied Physics Letters*, Vol. 107, 054101, 2015.

[61] Y. Alpert, and E. Jerby, "Coupled thermal-electromagnetic model for microwave heating of temperature-dependent dielectric media," *IEEE Trans. Plasma Science*, Vol. 27, pp. 555–562, 1999.

[62] U. Grosglick, V. Dikhtyar, and E. Jerby, "Coupled thermal-electromagnetic model for microwave drilling", JEE'02 Proceedings, pp. 146–151, European Symposium on Numerical Methods in Electromagnetics, March 6–8, 2002, Toulouse, France.

[63] F. P. Incropera, and D. P. Dewitt, *Fundamentals of Heat and Mass Transfer*, Wiley, New York, 1985.

[64] J. M. Catalá-Civera, A. J. Canós, P. Plaza-González, J. D. Gutiérrez, B. García-Baños, and F. L. Peñaranda-Foix, "Dynamic measurement of dielectric properties of materials at high temperature during microwave heating in a dual mode cylindrical cavity," *IEEE Trans. Microw. Theory Tech.*, Vol. 63, pp. 2905–2914, 2015.

[65] N. G. Evans, and M. G. Hamlyn, "Microwave firing at 915MHz: Efficiency and implications," *Mat. Res. Soc. Symp. Proc.*, Vol. 430, pp. 9–13, 1996.

[66] R. M. Hutcheon, J. Mouris, and C. A. Pickles, "Measurements of the complex dielectric constants of goethite and limonite minerals and nickel rich

laterite ores at temperature up to 1150K in Air for frequencies 912 MHz and 2.46GHz," Microwave Properties North, Deep River, Ontario, Canada, 1999.

[67] A. Birnboim, et al., "Comparative study of microwave sintering of Zinc oxide at 2.45, 30 and 83 GHz," *Jour. Amer. Ceram. Soc.*, Vol. 81, pp. 1493–1501, 1998.

[68] C. C. Goodson, "Simulation of microwave heating of mullite rods," M.Sc. Thesis, Virginia Tech., Dec. 1997; see also X. Wu, "Experimental and theoretical studies of microwave heating of thermal-runaway materials," Ph.D. Thesis, Virginia Tech, Dec. 2002 (both in http://scholar.lib.vt.edu/theses).

[69] U. Kolberg, and H. Roemer, "Microwave heating of glass," *Microwave: Theory and Application in Materials Processing*, V. D. E. Clark, J. G. P. Binner and D. A. Lewis (Eds.), Ceramic Transactions, Westerville, OH, Vol. 3, pp. 527–533, 2000.

[70] K. G. Ayappa, H. T. Davis, E. A. Davis, and J. Gordon, "Analysis of microwave heating of materials with temperature-dependent properties," *AIChE Journal*, Vol. 37, pp. 313–322, 2004.

[71] B. G. McConnell, "A coupled heat-transfer and electromagnetic model for simulating microwave heating of thin dielectric materials in a resonant cavity," M.Sc. Thesis, Virginia Poly. Inst., 1999.

[72] S. Soldatov, M. Umminger, A. Heinzel, G. Link, B. Lepers, and J. Jelonnek, "Dielectric characterization of concrete at high temperatures," *Cement and Concrete Composites*, Vol. 73, pp. 54–61, 2016.

[73] U. Schneider, "Concrete at high temperature: A general review," *Fire Safety Jour.*, Vol. 13, pp. 55–68, 1988.

[74] A. Taflove, and S. C. Hagness, *Computational Electrodynamics*, Artech House, Norwood MA, 2000.

[75] L. Ma, and D. L. Paul, "Experimental validation of combined electromagnetic and thermal FDTD model of microwave heating process," *IEEE Trans. Microwave Theory and Tech.*, Vol. 43, pp. 2565–2570, 1995.

[76] F. Torres, and B. Jecko, "Complete FDTD analysis of microwave heating process in frequency dependent and temperature dependent media," *IEEE Trans. Microwave Theory and Tech.*, Vol. 45, pp. 108–116, 1997.

[77] E. Jerby, Y. Meir, Y. Nerovny, O. Korin, R. Peleg, and Y. Shamir, "A silent microwave-drill for deep holes in concrete," *IEEE Trans. Microwave Theory and Techniques*, Vol. 66, pp. 522–529, 2018.

[78] R. W. P. King, and C. W. Harrison, *Antennas and Waves: A Modern Approach*, M.I.T. Press, Boston, MA, 1969.

[79] E. C. Jordan, and K. G. Balmain, *Electromagnetic Waves and Radiating Systems*, 2nd Edition, Prentice-Hall Inc., Inglewood Cliffs, NJ, 1968.

[80] Y. Meir, A. Salzberg, and E. Jerby, "Hotspot induced by low-power microwave drill: Transistor-based localized heaters and their new applications," Proc. Ampere 13th Int'l Conf., Toulouse, France, pp. 201–204, Sept. 5–8, 2011; also in. Proc. COMCAS-IEEE Int'l Conf., Tel Aviv, Israel, pp. 1–4, Nov. 7–9, 2011.

[81] V. Bilik, "Automatic impedance matching in high-power microwave applications," Proc. IMPI-48 Annual Symposium, June 18–20, 2014, New Orleans, pp. 6–9.

[82] G. Ong, and A. Akbarnezhad, *Microwave-Assisted Concrete Technology: Production, Demolition and Recycling*, CRC Press, Boca Raton, 2015.

[83] N. Makul, P. Rattanadecho, and D. K. Agrawal, "Applications of microwave energy in cement and concrete: A review," *Renewable Sustainable Energy Rev.*, Vol. 37, pp. 715–733, 2014.

[84] A. Buttress, A. Jones, and S. Kingman, "Microwave processing of cement and concrete materials: Towards an industrial reality?," *Cement and Concrete Research*, Vol. 68, pp. 112–123, 2015.

[85] N. Makul, B. Chatveera, and P. Ratanadecho, "Use of microwave energy for accelerated curing of concrete: A review," *Songklanakarin J. Sci. Technol.*, Vol. 31, pp. 1–13, 2009.

[86] R. V. Silva, J. de Brito, R. K. Dhir, "Properties and composition of recycled aggregates from construction and demolition waste suitable for concrete production," *Constr. Build. Mater.*, Vol. 65, pp. 201–217, 2014.

[87] A. Akbarnezhad, K. S. C. Kuang, and K. C. G. Ong, "Temperature sensing in microwave heating of concrete using fibre Bragg grating sensors," *Mag. Concr. Res.*, Vol. 63, pp. 275–285, 2011.

[88] E. Jerby, Y. Shamir, R. Peleg, and Y. Aharoni, "A silent mechanically-assisted microwave-drill for concrete with integrated adaptive impedance matching," Proc. AMPERE-14 Int'l Conf. Microwave & RF Heat., Nottingham, UK, Sept. 16–19, 2013, pp. 267–270 (Unpublished).

[89] Y. Shoshani, T. Levin, and E. Jerby, "Concrete cutting by a solid-state, localized microwave-heating (LMH) applicator," *Trans. IEEE-MTT*, Vol. 69, pp. 4237–4245, 2021.

[90] IEEE Standard C95.1, *IEEE standard for safety levels with respect to human exposure to radio frequency electromagnetic fields, 3 kHz to 300 GHz*, Rev. 2005.

[91] S. Saha, S. Pal, and J. A. Albright, "Surgical drilling: Design and performance of an improved drill," *J. Biomech. Eng.*, Vol. 109, pp. 245–252, 1982.

[92] J. T. Payne, G. M. Peavy, L. Reinisch, and D. C. Van Sickle, "Cortical bone healing following laser osteotomy using 6.1 μm wavelength," *Lasers in Surgery and Medicine*, Vol. 29, pp. 38–43, 2001.

[93] M. Mashevich, D. Folkman, A. Kesar, A. Barbul, R. Korenstein, E. Jerby, and L. Avivi, "Exposure of human peripheral blood lymphocytes to electromagnetic fields associated with cellular phones leads to chromosomal instability," *Bioelectromagnetics*, Vol. 24, pp. 82–90, 2003.

[94] G. S. Gazelle, S. N. Goldberg, L. Solbiati, and T. Livraghi, "Tumor ablation with radio-frequency energy", *Radiology*, Vol. 217, pp. 633–646, 2000.

[95] M. R. Callstrom, J. W. Charboneau, M. P. Goetz, J. Rubin, G. Y. Wong, et al., "Painful metastases involving bone: feasibility of percutaneous CT- and US- guided radio-frequency ablation," *Radiology*, Vol. 224, pp. 87–97, 2002.

[96] M. Fugenfirov, Y. Meir, A. Shelef, Y. Nerovny, E. Aharoni, and E. Jerby, "Incremental solidification (toward 3D-printing) of magnetically-confined metal-powder by localized microwave heating", *COMPEL—Int'l Jour. Comp. Math. Elect. Eng.*, Vol. 37, pp. 1918–1932, 2018.

[97] A. Shelef, and E. Jerby, "Incremental solidification (toward 3D-printing) of metal powders by transistor-based microwave applicator," *Materials and Design*, Vol. 185, Art. No. 108234, 2020.

[98]   Y. Meir, and E. Jerby, "Breakdown spectroscopy induced by localized micro-waves for material identification," *Microw. Opt. Technol. Lett.*, Vol. 53, pp. 2281–2283, 2011.

[99]   O. Hammi, and F. M. Ghannouchi, "Comparative study of recent advances in power amplification devices and circuits for wireless communication infrastructure," 16th IEEE Int'l Conf. on Electronics, Circuits, and Systems, ICECS'2009, Tunisia, pp. 379–382, 2009.

[100] H. Blanck, J. R. Thorpe, R. Behtash, J. Splettstößer, P. Brückner, S. Heckmann, H. Jung, K. Riepe, F. Bourgeois, M. Hosch, D. Köhn, H. Stieglauer, D. Flo-riot, B. Lambert, L. Favede, Z. Ouarch, and M. Camiade, "Industrial GaN FET technology," *Int'l Jour. Microwave and Wireless Technologies*, Vol. 2, pp. 21–32, 2010.

[101] G. Chen, Z. Chen, S. Hai, P. Wu, Z. Li, and Z. Feng, "Microwave power of S-band 20mm SiC MESFETs," IEEE Int'l Conf. Electron Devices and Solid-State Circuits, EDSSC 2009, pp. 484–486, 2009.

[102] O. Latry, P. Dherbécourta, K. Mourguesa, H. Maananeb, J. P. Sipmab, F. Cor-nub, P. Eudelineb, and M. Masmoudic, "A 5000 h RF life test on 330-W RF-LDMOS transistors for radars applications," *Microelectronics Reliability*, Vol. 50, pp. 1574–1576, 2010.

[103] D. L. Gershon, J. P. Calame, Y. Carmel, T. M. Antonsen, and R. M. Hutcheon, "Open-ended coaxial probe for high-temperature and broad-band dielectric measurements," *IEEE Trans. Microwave Theory and Tech.*, Vol.47, pp. 1640–1648, 1999.

Chapter 4

Microwave Welding as an Alternative Nonconventional Welding Technique

Guru Prakash, Ramkishor Anant, Rahul Gupta, and Amit Bansal

Contents

4.1 Introduction

Fusion welding is one of the most commonly used process; it is also largely used for joining dissimilar metals. For the versatile application in general weld fabrication of metallic sections of stainless steel, aluminum, cast iron, and others, there may be several choices of fusion welding processes, which can be divided into two categories: (1) conventional techniques such as shielded metal arc, gas tungsten arc, gas metal arc and submerged arc welding and (2) nonconventional techniques such as electron beam, ultrasonic, laser beam, friction welding and microwave welding (Figure 4.1). Every process has a different weld thermal cycle depending on rate of weld deposition, welding parameters and shielding environment. The amount of weld deposition, which also considerably influences the severity of thermal characteristics of a weld, can be considerably reduced by using a microwave welding technique. Among all these conventional and nonconventional welding processes, the use of microwave welding is gaining more attention in the fabrication of weld joint, especially in the case of dissimilar welding of stainless steel to carbon steel, due to their capability to produce a superior quality weld in comparison to that of the commonly used gas metal arc welding, gas tungsten arc welding, shielded metal arc welding and submerged arc welding, with improved mechanical properties (tensile strength, hardness) and metallurgical properties in terms of as heat-affected zone (HAZ) and so on. In metal joining, a comparatively low heat input welding process is generally preferred for a good quality weld to improve mechanical and metallurgical

DOI: 10.1201/9781003248743-4

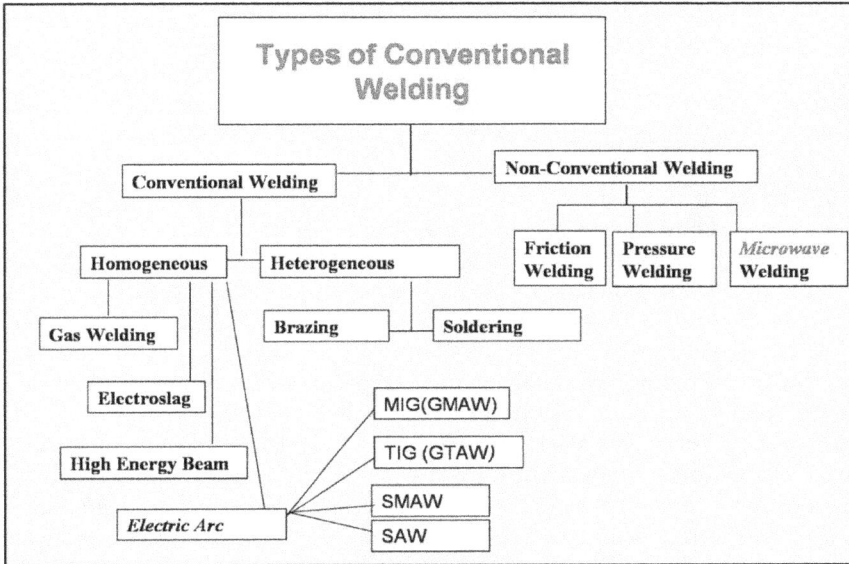

Figure 4.1 Classification of welding processes.

properties of the joints. In this way, the superiority of the microwave weld over the other welds is primarily understood by its ability to produce comparatively cleaner weld, improved mechanical properties, better integrity and higher economy. But the critical control of heating characteristics and behavior of metal joining as a function of welding parameters often makes the application of the microwave welding process relatively complicated to achieve the desired weld quality, especially with respect to defect and metallurgical characteristics of the weld.

The energy input involved in a microwave welding process, causing the partial melting of the base metals and the diffusion of powders, also develops certain reactions in a very short period, which results in some dramatic changes in the microstructure of HAZ near the fusion line. These changes primarily result from the rapid heating and cooling of the base materials during welding. In order to control these changes, it is necessary to regulate the thermal behavior of the welding process, especially in reference to the critical durations of the maximum temperature and the temperature above some critical range, as well as cooling rate prevailing in the interactive locations of the materials involved with respect to diffusion of active elements, phase transformations and development of residual stress at the joint mainly during the dissimilar welding process.

The dissimilar metal welding of metallic sections is a bit critical. The difficulties primarily include problems largely associated with difference in coefficient of thermal expansion (CTE) and the thermal conductivity of the material to be joined. The large differences in CTE and thermal conductivity causes the undesirable development and distribution of residual stresses, development of undesirable weld chemistry due to dilution, metallurgical incompatibility primarily with respect to the formation of undesirable phases in the weld as well as HAZ and the segregation of high and low melting phases due to chemical mismatch [1–3]. These problems can be overcome by the microwave welding process due to the heat concentration projected in a very small area. The historical perspectives of the major developments in welding technologies are shown in Figure 4.2 [4].

Although there exists a fairly small number of publications on the joining of similar and dissimilar metals by the microwave welding process, a review in the area shows that majority of this literature is mainly concentrated on only process development as this process is still under development. As such, there are few data available in the open literature regarding other mechanical and metallurgical properties. A systematic study from a scientific angle is therefore desirable in this field for effective usage of this microwave weld joint technique.

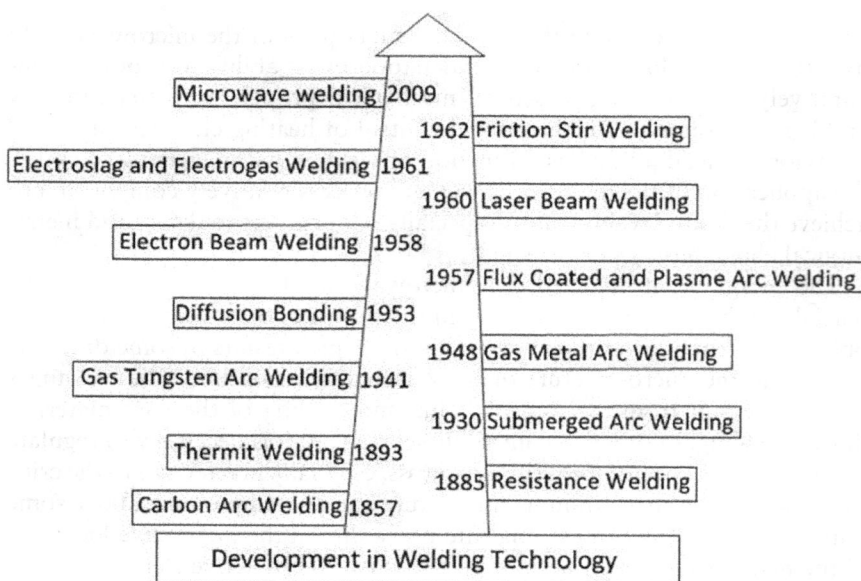

Figure 4.2 Historical perspective in the development of welding technology.

In recent years, the interest of welding engineers is largely being focused on the use of the microwave welding process. The inherent advantages associated with this process are like economy with improved weld quality, better mechanical properties and good performance in welding. The use of the microwave welding process has also been found to further improve the weld quality with respect to microstructure and other characteristics of weld and HAZ as compared to other conventional techniques.

4.2 Materials Processed by Microwave Technique

Today, the microwave processing of materials is gaining popularity. Microwave energy has been widely applied to processing metals, ceramics and composite materials like polymer matrix composites, metal matrix composites and ceramic matrix composites, among others. By using microwave radiation at frequency 2.45 GHz, nearly full density of alumina has been achieved when it is sintered at 1350°C for 50 min and with conventional heating at 62% density is achieved when sintered at the same temperature [5]. By using microwaves, many commercial powders and their alloys can be sintered. It has been reported that sintering powders that contain Gr (0.8%), Cu (2%) and Fe has been done in a microwave field at 1200°C for 30 min with good density [5]. It has also been reported that metal powder of cobalt (Co) has been microwave sintered in pure H2 environment at different temperatures ranging from 900–1200°C with 1 atm pressure for 10 min. At 900°C, the reported density was 8700 kg/m^3, while at 1000–1050°C, it was 8880 kg/m^3 and at 1100–1200°C theoretical density was 8890 kg/m^3. Rodiger et al. [6] had used a 2.45-GHz microwave to sinter hard metals, where sintering temperature (1300°C) was attained in 1.5 h with 1 kW power, whereas with the conventional process, the same temperature was attained in 5 h with 4.5 kW. Prabhu et al. [7] used microwaves to compare the sinterability of as-received powder with activated tungsten (W) powder. It was found that a better densification was obtained with W powder due to its higher specific energy and small particle size. The rapid sintering of Mg, Al and Pb free solder by using the two-direction microwave was reported by Gupta et al. [8]. They observed that the density of both the samples, one microwave-sintered and conventionally sintered, were the same but that microwave-sintered Al and Mg showed excellent ultimate tensile strength (UTS) with minimal increase in microhardness [8].

Chiu et al. [9] reported cladding of NiTi on AISI 316 stainless steel substrates by a microwave-assisted brazing process. Later, the cladding of metallic powder on a metallic substrate and composite powder on a metallic substrate by microwave hybrid heating (MHH) process was successfully reported by Gupta and Sharma in 2011 [10, 11]. The microwave joining of thin metallic steel sheet in the thickness range of 0.1–0.3 mm was also reported [12]. The authors showed that the localized arcing was enough to melt such thin test sheets by using a 2-kW multimode magnetron. Bulk metallic materials, however, could not be joined by this process due to the reflection of microwaves by metals as the spark produced was large, which would be harmful to the waveguide and the magnetron. The possible solution of joining metallic materials was reported by Sharma et al. [13] in the form of patent. Srinath et al. [14, 15, 16], in 2011, reported joining similar and dissimilar bulk metals in atmospheric conditions with the help of the MHH technique. The applications of microwaves in processing metallic materials in different forms are shown in Figure 4.3.

In material processing, microwave joining can be used for a wide variety of materials. The versatile process can be used for similar material as well as non-similar materials. The following section highlights in brief the work done by various researchers to carry out joining different materials utilizing the cleaner source of microwave energy.

Microwave joining requires many inputs, which are shown in Figure 4.4. In microwave joining, selective heating and rapid heating can be achieved [17]. In microwave joining, both direct heating and indirect heating methods were utilized [18]. Microwaves cannot be directly applied to metal/alloys due to reflection of the metals to the incident microwave radiation. In order to join metal/alloys, it needs to be covered by microwave absorber material, which is known as a susceptor. The heating happens in three stages. In the first stage, the temperature rise of the susceptor material takes place by absorbing microwave energy at room temperature. In the second stage, the heat from the hot susceptor is transferred to the target materials to be joined in a conventional mode of heat transfer. Subsequently, the role of the separator comes in the picture. The separator separates the susceptor and the target material once the target achieves the critical temperature [19]. Once the target attains the critical temperature, then it is directly heated by microwave, which leads to rapid heating [20]. Figure 4.4 shows the various input

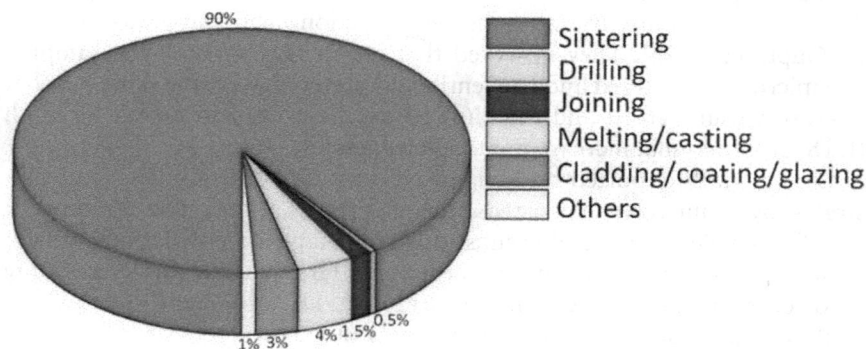

Figure 4.3 Application of microwave in processing metallic materials.

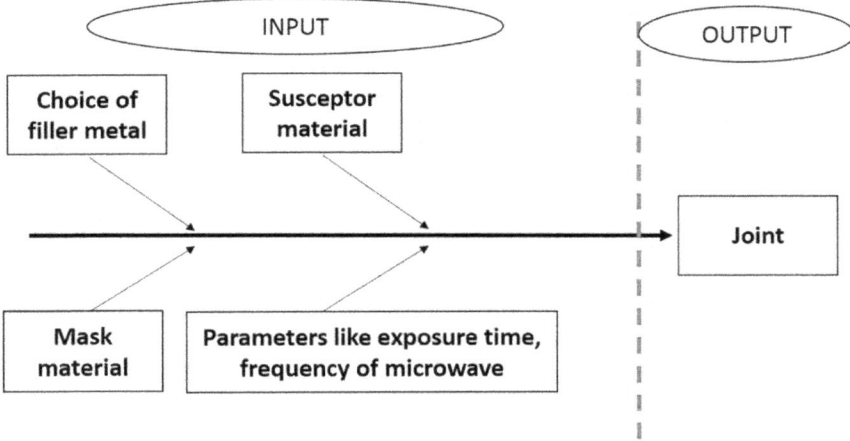

Figure 4.4 Different inputs for microwave joining.

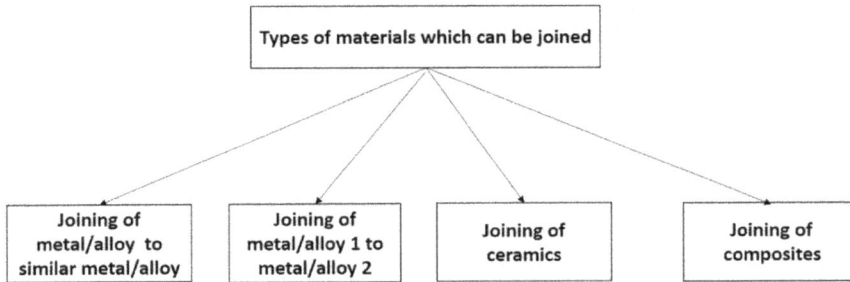

Figure 4.5 Different types of materials joined by microwave joining.

parameter required for joining materials utilizing microwaves as a source of energy.

Furthermore, the various similar and dissimilar metals and alloys can be joined by microwave joining techniques. Microwave joining can also be used for joining composites. Figure 4.5 illustrates the different materials that can be joined by using this microwave energy. Microwaves can be used to attain temperatures as high as 1400°C. Microwaves can also be used for surface engineering to enhance the surface properties of metal/alloys [21–23].

4.3 Applications of Microwave for Joining Materials

Table 4.1 list different type of material joined by microwave joining technique. Some cases of microwave joining with features are discussed in the

Table 4.1 Microwave Joining Different Materials

Sr. No.	Authors	Materials Joined	Type of Joint	Interlayer Powder	Process Parameter	Findings/Observations
1	Pal et al., 2020 [43]	SS304 and SS316	Dissimilar	Ni	Microwave radiation having a frequency of 2.45 GHz and a process time of 360, 380, 400 and 420 s was used.	The joint has elements like Fe, C, Ni and Cr, which implies that optimum melting of Ni filler metal and stainless steels have taken place due to microwave heating.
2	Pal et al., 2020 [44]	SS304 and SS316	Dissimilar	Ni	Microwave radiation having a frequency of 2.45 GHz and a process time of 360, 380, 400 and 420 s was used.	Microstructure reveals that joint has a dense microstructure with negligible porosity and hence has higher hardness and tensile strength in comparison to parent alloys
3	Kumar et al., 2020 [45]	SS304 and SS316	Dissimilar	Powder not used	Microwave oven of 900 W and a frequency of 2.45 GHz was used. The exposure time ranged from 410–460 s with an interval of 10 s.	Strong joint having high strength and hardness was achieved by microwave joining.
4	Kumar and Sehgal, 2020 [46]	SS2205 with SS2205	Similar	Powder not used	Microwave oven of 900 W and a frequency of 2.45 GHz was used. The exposure time was 240–300 sec with an increment of 30 sec.	The hardness of the joint was 63% higher than of the base metal.
5	Tamang and Aravindam, 2019 [47]	Boron Nitride and WC-Co	Dissimilar	Braze alloy having Ag, Cu, In and Ti	Microwave oven of 700 W and a frequency of 2.45 GHz was used. The exposure time was 10–13 min with an increment of 30 s.	WC-Co and cBN were successfully joined by microwave joining. The joint exhibited good strength than that of the parent material.
6	Kumar et al., 2020 [48]	SS304 and SS316	Dissimilar	SS316	Microwave oven of 900 W and a frequency of 2.45 GHz was used. The exposure time was 360 sec.	The joint reported 72% higher hardness than that of base alloy. Strong metallurgical bonding was achieved in the joint
7	Bagha et al., 2019 [49]	SS304 and SS304	Similar	Powder not used	Microwave oven of 800 W and a frequency of 2.45 GHz was used.	The joint had higher hardness than that of base metal. The ultimate tensile strength was 271 MPa for the joint.
8	Tamang and Aravindan (2019) [50]	Copper and SS304	Dissimilar	Ni	Microwave oven of 700 W and a frequency of 2.45 GHz was used.	Strong metallurgical bonding was achieved by microwave joining.

#	Reference	Materials	Similar/Dissimilar	Filler	Process parameters	Remarks
9	Soni et al., 2018 [51]	SS316 and SS316	Similar	Ni	Microwave oven of 800 W and frequency of 2.45 GHz was used.	Use of nanofiller powder enhances the tensile strength by 24% in comparison to micro filler powder.
10	Badiger et al. 2018 [52]	Inconel 625 and Inconel 625	Similar	Ni	Microwave oven of 900 W and frequency of 2.45 GHz was used.	Higher strength of joint is achieved by use of finer powder.
11	Singh et al., 2018 [53]	Cast iron and cast iron	Similar	Ni	Microwave oven of 900 W and a frequency of 2.45 GHz was used. Exposure time was 240–480 s with an interval of 40 s.	The joint had a hardness of 185 VHN and tensile strength of 196 MPa which is good in term of application.
12	Gamit et. al., 2021 [24]	MS Pipes	Similar	Ni	Microwave oven of 900 W and a frequency of 2.45 GHz was used.	Micro indentation hardness of the joint zone was observed to be higher (572 HV) than the hardness of the base metal (397 HV) owing to formation of the Ni-based intermetalics and carbides.
13	Bajpai et. al., 2012 [25]	Grewiaoptivafiber mat (GOF) and Nettle fiber mat (NF)	Dissimilar	charcoal	Microwave oven of 900 W and a frequency of 2.45 GHz was used.	Microwave joining provides higher joint strength as compared to adhesive bonding
14	Samyal et. al., 2021 [26]	SS202 and SS202	Similar	Ni	Microwave oven of 700 W and a frequency of 2.45 GHz with a capacity of 25L was used. The exposure time varied between 15–25 min.	With an exposure of 19 min, the best joining has been obtained.
15	Srinath et al., 2011b [15]	Mild steel and stainless steel	Dissimilar	Ni powder	Microwave oven of 900 W and frequency of 2.45 GHz was used. The exposure time was 600 s.	The joint had a hardness of 130VHN and ultimate tensile strength of 340.16 MPa, which is good in terms of application.
16	Bansal et al., 2015 [31]	Inconel 718 and Austenitic stainless steel (SS-316 L)	Dissimilar	Inconel 718 powder	Microwave oven of 900 W and a frequency of 2.45 GHz was used. The exposure time was 600 s.	The joint had a hardness of 230 ± 5 VHN and ultimate tensile strength of 517.5 MPa, which is good in term of application. The failure of the weld was mainly mixed mode.

(Continued)

Table 4.1 (Continued)

Sr. No.	Authors	Materials Joined	Type of Joint	Interlayer Powder	Process Parameter	Findings/Observations
17	Badiger et. al., 2015 [35]	Inconel 625	Similar	Ni powder	Microwave oven of 900 W and a frequency of 2.45 GHz was used. The exposure time was 21 min.	The hardness and tensile strength of the welded joint is 350 ± 5 Hv and 326 MPa. The failure of the joint is mixed mode failure.
18	Pal et. al., 2020 [35]	SS304 and SS316	Dissimilar	Ni powder	Microwave oven of 800 W and a frequency of 2.45 GHz was used. For joining, the time is varied from 270–480 s in the steps of 30 s.	Both micro-hardness and micro-tensile strength increase significantly with decreasing Ni particle size
19	Bagha et. al., 2017 [54]	SS304 with SS304	Similar	Ni	Microwave oven of 800 W and a frequency of 2.45 GHz with a capacity of 23L was used.	Size of Ni powder decides physical and mechanical properties.
20	Singh et. al., 2015 [55]	Al plates	Similar	Al powder	Microwave oven of 900 W and a frequency of 2.45 GHz was used. The exposure time was 600 s.	Uniform microstructure with good mechanical bonding was obtained.
21	Bansal et. al., 2014 [56]	SS316 with SS316	Similar	SS316	Microwave oven of 900 W and a frequency of 2.45 GHz was used.	Microhardness of the joint was significantly higher than the base. Porosity at the joint was also low. The joint showed ductile fracture.
22	Dwivedi and Sharma, 2014 [57]	1018 Mild Steel	Similar	Ni powder	Three different microwave ovens with A rated power of 800, 850, and 900 W and fixed frequency of 2.45 GHz were used.	Tensile strength increases with decreasing rated power output. Optimum tensile strength was obtained with 800-W power output, 1000 s, and 800 °C.

following. Gamit et al. [24] have reported the joining of mild steel pipes by microwave joining. A microwave with 2.45 GHz generated with 900 W was used in the technique. Ni powder was used as interlayer powder as a sandwich between the mild steel. It is reported that microwave joining is not only an eco-friendly process but also less time-consuming process. The joint zone has higher hardness of 572 VHN in comparison to the base metal, which has a higher hardness of 397 VHN.

Bajpai et al. [25] have reported that microwave joining can be used for joining composites. Natural fiber–based composites can be successfully joined by microwave joining. It was observed that the joint formed by microwave joining is far superior to the joint formed by adhesive joining.

Samyal et al. [26] have joined SS202 sheets by microwave joining. Ni was used as the interlayer material for joining the sheet.

A new nonconventional technique that uses microwaves at a frequency of 2.45 GHz for joining is hybrid carbon microwave joining (HCMJ). There are many advantages of HCMJ technique, such as less utilization of powder, a better hardness, volumetric heating, eco-friendly, reduced HAZ and good tensile strength [27]. The better mechanical properties were observed with HCMJ joints of pure Ni powder than EWAC-based counterparts [28].

MHH is a new process for joining bulk metals. This process is fast and economical and offers selective heating without any constrain in joint configurations [29]. Hence, a domestic microwaves oven can be used to join dissimilar metals [30–31]. The product sintered by microwaves showed lesser porosities and superior properties than the product sintered by conventional method [32]. It has been observed that MHH has been recently used to examine joining of polymers [33] and bulk metals [34–35].

To weld nickel-based alloys, conventional welding processes are easily available and offer more flexibility, but the large amount of heat involved not only deteriorates the chemical and mechanical properties but also reduces corrosion resistance due to which the microstructures of the base metal and the fusion zone are significantly changed [36, 37]. Furthermore, conventional welding processes demand high power, which also needs to be addressed. Thus, the MHH technique is used to join Inconel-625 alloy. In microwave heating, energy in the form of electromagnetic waves is directly converted into heat energy, which is very fast, and volumetric heating takes place, whereas in conventional heating, the thermal conductivity phenomenon is responsible for transferring of energy [38].

There is another joining technique which uses microwave is microwave-assisted combustion synthesis (MACS). In this technique, it is expected that microwaves are absorbed by both SiC substrates and reactive powders. The tendency of absorption depends on the dielectric and the material's electrical properties. This not only enhances chemical bonding and inter-diffusion but also reduces the HAZ [39].

MACS was successfully used to join SiC samples where an exothermic reaction between graphite powders, titanium and silicon, mixed in a 1:1:1 atomic ratio, was used. The main advantages of this technique are selective absorption of the microwaves by the joining area, the high speed and the low joining pressure [39].

Meek and Blake [40] used multimode microwave oven to join alumina-alumina ceramic by using commercial sealing glass interlayers. The strength of the joint was comparable with other high-temperature joining methods, such as brazing. Palaith and Silberglitt [41] used microwave at a frequency of 2.45 GHz to join oxide and nonoxide ceramics. Similar findings have been reported by Fukushima et al. [42] by using microwave at a frequency of 6 GHz. Table 4.1 list the different type of material joined by utilizing this nonconventional source of energy in the form of the MHH technique.

4.4 Conclusion

A wide range of similar and dissimilar materials can be joined by the microwave joining technique. Even composites and ceramics can be joined using this technique.

The advantage of microwave joining is as follows:

1. Wide range of similar and dissimilar materials can be joined by microwave joining.
2. Smaller HAZ is there in the joint as selective and volumetric heating is achieved.
3. Time consumed is low due to heating takes place at the atomic and molecular levels.
4. Environmentally friendly process as it generates less pollution.

The limitations associated with this technique are as follows:

1. Small sized specimen can only be by microwave joining technique. It is a challenge to join large-sized metal/alloys.
2. Availability and knowledge of the susceptor of the material are important.
3. Porosity in the joint obtained from microwave joining is a concern.
4. Microwave joining is an evolving and novel technology and requires standardization of the process. Researchers are continuously working on it to refine the process.

References

[1] Satnam Singh, Praveen Singh, Dheeraj Gupta, Vivek Jain, Rohit Kumar, Sarbjeet Kaushal, Development and characterization of microwave processed cast

iron joint, *Engineering Science and Technology, an International Journal*, 22(2), (2019), 569–577.

[2] M. Ebrahimnia, F.M. Ghaini, S. Gholizade, M. Salari, Effect of cooling rate and powder characteristics on the soundness of heat affected zone in powder welding of ductile cast iron, *Mater. Des.*, 33 (2012), 551–556.

[3] H.S. Ku, E. Siores, J.A.R. Ball, Review: Microwave processing of materials: Part II, *HKIE Trans.*, 8 (2001), 38–43, 10.

[4] M.M. Schwartz, *Metal joining manual*, McGraw-Hill Book Company, New York, 1979.

[5] Kristen H. Brosnan, Gary L. Messing, D.K. Agrawal, Microwave sintering of alumina at 2.45 GHz, *J Am CerSoc*, 86 (8), (2003), 1307–1312.

[6] K. Rodiger, K. Dreyer, T. Gerdes, M. PoradaWillert, Microwave sintering of hard metals, *Int J Refract Met Hards Mater*, 16 (1998), 409–416.

[7] G Prabhu, Chakraborty Amitava, Sarma Bijoy, Microwave sintering of tungsten, *Int. J Met Hard Mater* 27 (2009), 545–548.

[8] P. Gupta, S. Kumar, A. Kumar, Study of joint formed by tungsten carbide bearing alloy through microwave welding, *Mater Manuf Process*, 28(5), (2013), 601–604.

[9] K.Y. Chiu, F.T. Cheng, H.C. Man, A preliminary study of cladding steel with NiTi by microwave-assisted brazing, *Materials Science and Engineering: A*, 407 (2005), 273–281.

[10] D. Gupta, A.K. Sharma, Investigation on sliding wear performance of WC10Co2Ni cladding developed through microwave irradiation, *Wear*, 271 (2011a), 1642–1650.

[11] D. Gupta, A.K. Sharma, Development and microstructural characterization of microwave cladding on austenitic stainless steel, *Surface and Coatings Technology*, 205 (2011b), 5147–5155.

[12] E. Siores, D. Rego, Microwave applications in material joining, *Journal of Materials Processing Technology*, 48 (1995), 619–625.

[13] A.K. Sharma, M.S. Srinath, P. Kumar, Microwave joining of metallic materials, Indian Patent, Application No. 1994/Del/2009, 2009.

[14] M.S. Srinath, A.K. Sharma, P. Kumar, A new approach to joining of bulk copper using microwave energy, *Materials and Design*, 32 (2011a), 2685–2694.

[15] M.S. Srinath, A.K. Sharma, P. Kumar, Investigation on microstructural and mechanical properties of microwave processed dissimilar joints, *Journal of Manufacturing Processes*, 13 (2011b), 141–146.

[16] M.S. Srinath, A.K. Sharma, P. Kumar, A novel route for joining of austenitic stainless steel (SS-316) using microwave energy, *Proceedings of the Institution of Mechanical Engineers Part B, Journal of Engineering Manufacture*, 225 (2011c), 1083–1091.

[17] S. Singh, D. Gupta, V. Jain, Recent applications of microwaves in materials joining and surface coatings, *Proc. Inst. Mech. Eng. Part B J. Eng. Manuf.*, 230 (2016), 603–617.

[18] R.R. Mishra, A.K. Sharma, Microwave–material interaction phenomena: Heating mechanisms, challenges and opportunities in material processing, *Compos. Part A Appl. Sci. Manuf.*, 81 (2016), 78–97.

[19] R.R. Mishra, A.K. Sharma, **A review of research trends in microwave processing of metal-based materials and opportunities in microwave metal casting**, *Crit. Rev. Solid State Mater. Sci.*, 41 (2016), 217–255.

[20] M. Gupta, W.L.E. Wong, Enhancing overall mechanical performance of metallic materials using two-directional microwave assisted rapid sintering, *Scr. Mater.*, 52 (2005), 479–483.

[21] M.S. Srinath, A.K. Sharma, P. Kumar, A novel route for joining of austenitic stainless steel (SS-316) using microwave energy, *Proc. Inst. Mech. Eng. Part B J. Eng.*, 225 (2011), 1083–1091.

[22] K.Y. Chiu, F.T. Cheng, H.C. Man, A preliminary study of cladding steel with NiTi by microwave-assisted brazing, *Mater. Sci. Eng. A*, 407 (2005), 273–281.

[23] **R.R. Mishra, A.K. Sharma, On mechanism of in-situ microwave casting of aluminium alloy 7039 and cast microstructure**, *Mater. Des.*, 112 (2016), 97–106.

[24] D. Gamit, R.R. Mishra, A.K. Sharma, Joining of mild steel pipes using microwave hybrid heating at 2.45 GHz and joint characterization, *J Manuf Process*, 27 (2017), 158–168.

[25] P.K. Bajpai, I. Singh, J. Madaan, Joining of natural fiber reinforced composites using microwave energy: Experimental and finite element study, *Mater Des*, 35 (2012), 596–602.

[26] R.Samyal, A.K. Bagha, R. Bedi, Evaluation of modal characteristics of SS202-SS202 lap joint produced using selective microwave hybrid heating, *J Manuf Process*, 68 (2021), 1–13.

[27] M. Gupta, W.L.E. Wong, Enhancing overall mechanical performance of metallic materials using two-directional microwave assisted rapid sintering, *Scripta Mater*, 52 (2005), 479–483.

[28] L. Bagha, S. Sehgal, A. Thakur, Comparative analysis of microwave-based joining/welding of SS304-SS304 using different interfacing materials. *MATEC Web Conf*, 57 (3001), (2016), 1–4.

[29] E. Siores, D.D. Rego, Microwave applications in materials joining, *J. Mater. Process. Technol.*, 48 (1995), 619–625.

[30] M.S. Srinath, A.K. Sharma, P. Kumar, Investigation on microstructural and mechanical properties of microwave processed dissimilar joints, *J. Manuf. Process.*, 13 (2011), 141–146.

[31] A. Bansal, A.K. Sharma, S. Das, P. Kumar, On microstructure and strength properties of microwave welded Inconel 718/stainless steel (SS-316L), *Proc. Inst. Mech. Eng. Part L J. Mater. Des. Appl.*, (2015), 1–10.

[32] P. Yadoji, R. Peelamedu, D. Agrawal, R. Roy, Microwave sintering of Ni-Zn ferrites: Comparison with conventional sintering, *Mater. Sci. Eng. B Solid-State Mater. Adv. Technol.*, 98 (2003), 269–278.

[33] S. Grohmann, G. Langhans, A. Reindl, V. Sidarava, M.F. Zaeh, Investigation of reactive bimetallic Ni-Al particles as a heat source for microwave-assisted joining, *J. Mater. Process. Technol.*, 282 (2020), 116637.

[34] M.S. Srinath, A.K. Sharma, P. Kumar, A new approach to joining of bulk copper using microwave energy, *Mater. Des.*, 32 (2011), 2685–2694.

[35] R.I. Badiger, S. Narendranath, M.S. Srinath, Joining of inconel-625 alloy through microwave hybrid heating and its characterization, *J. Manuf. Process.*, 18 (2015), 117–123.

[36] K.H. Song, W.Y. Kim, K. Nakata, Evaluation of microstructures and mechanical properties of friction stir welded lap joints of Inconel 600/SS-400, *Mater Des*, 35 (2012), 126–132.

[37] H.R. Zareie Rajani, S.A.A. Akbari Mousavi, The effect of explosive welding parameters on metallurgical and mechanical interfacial features of Inconel 625/plain carbon steel bimetal plate. *Mater SciEng A*, 556 (2012), 454–464.

[38] Oghbaei M, Mirzaee O, Microwave versus conventional sintering: A review of fundamentals, advantages and applications, *J Alloys Compd*, 494 (2010), 175–189.

[39] R. Rosa, P. Veronesi, S. Han, V. Casalegno, M. Salvo, E. Colombini, C. Leonelli, M. Ferraris, Microwave assisted combustion synthesis in the system Ti-Si-C for the joining of SiC: Experimental and numerical simulation results, *J. European Ceramic Society*, 33 (2013), 1707–1719.

[40] T.T. Meek, R.D. Blake, Ceramic-ceramic seals by icrowave heating, *J. Mater. Sci. Lett.*, 5 (1986), 270–274.

[41] D. Palaith, R. Silberglitt, Microwave joining of ceramics, *Ceramic Bulletin*, 68(9), *American Ceramic Society, Westerville*, (1989), 1601–1606.

[42] H. Fukushima, T. Yamanaka, M. Matsui, Microwave heating of ceramics and its applications to joining, *J. Mater. Res.*, 5(2), (1990), 397–405.

[43] M. Pal, S. Sehgal, H. Kumar, D. Goyal, Use of nickel filler powder in joining SS304-SS316 through microwave hybrid heating technique, *Met. Powder Rep.* (2020), 1–6.

[44] M. Pal, V. Kumar, S. Sehgal, H. Kumar, K. Kuldeep, A.K. Bagha, Microwave hybrid heating based optimized joining of SS304/SS316, *Mater. Manuf. Process.*, 36 (2021), 1–7.

[45] A. Kumar, S. Sehgal, S. Singh, A.K. Bagha, Joining of SS304-SS316 through novel microwave hybrid heating technique without filler material, *Mater. Today Proc.* (2020).

[46] V. Kumar, S. Sehgal, Joining of duplex stainless steel through selective microwave hybrid heating technique without using filler material, *Mater. Today Proc.* 28 (2020), 1314–1318.

[47] S. Tamang, S. Aravindan, Brazing of cBN to WC-Co by Ag-Cu-In-Ti alloy through microwave hybrid heating for cutting tool application, *Mater. Lett.* 254 (2019), 145–148.

[48] S. Kumar, S. Sehgal, S. Singh, A.K. Bagha, Investigations on material characterization of joints produced using microwave hybrid heating, *Mater. Today Proc.* (2020).

[49] L. Bagha, S. Sehgal, A. Thakur, H. Kumar, D. Goyal, Low cost joining of SS304-SS304 through microwave hybrid heating without filler-powder, *Eng. Res. Express*, 1 (2019).

[50] S. Tamang, S. Aravindan, Joining of Cu to SS304 by microwave hybrid heating with Ni as interlayer, En AMPERE 2019, 17th Int. Conf. Microw. High Freq. Heat., 2019, pp. 98–104.

[51] P. Soni, S. Sehgal, H. Kumar, A.P. Singh, Effect of Nickel nano-powder on joining SS316-SS316 through microwave hybrid heating, *Adv. Mater. Manuf. Chararcterization*, 8 (2018), 44–48.

[52] R.I. Badiger, S. Narendranath, M.S. Srinath, Optimization of process parameters by taguchi grey relational analysis in joining inconel-625 through microwave hybrid heating, *Metallogr. Microstruct. Anal.*, 8 (2019), 92–108.

[53] S. Singh, P. Singh, D. Gupta, V. Jain, R. Kumar, S. Kaushal, Development and characterization of microwave processed cast iron joint, *Eng. Sci. Technol. an Int. J.*, 22 (2018), 569–577.

[54] L. Bagha, S. Sehgal, A. Thakur, H. Kumar, Effects of powder size of interface material on selective hybrid carbon microwave joining of SS304–SS304, *J. Manuf. Process.*, 25 (2017), 290–295.

[55] S. Singh, N.M. Suri, R.M. Belokar, Characterization of joint developed by fusion of aluminum metal powder through microwave hybrid heating, *Mater. Today Proc.*, 2 (2015), 1340–1346.

[56] A. Bansal, A.K. Sharma, S. Das, P. Kumar, On microstructure and strength properties of microwave welded Inconel 718/ stainless steel (SS-316L), *Proc. Inst. Mech. Eng. Part L J. Mater. Des. Appl.*, 230 (2015), 939–948.

[57] S.P. Dwivedi, S. Sharma, Effect of process parameters on tensile strength of 1018 mild steel joints fabricated by microwave welding, *Metallogr. Microstruct. Anal.*, 3 (2014), 58–69.

Chapter 5

Clad Developments Through Microwave Hybrid Heating Technique

Processing and Properties

Gudala Suresh, M.R. Ramesh, Ajit M. Hebbale, and M.S. Srinath

Contents

5.1 Introduction

5.1.1 Background

The demand for better quality processes with lower production/process costs is increasing in the industry. One such approach is to use microwave heating. Earlier, in the period between 1950 and 1970, microwaves were mainly used to heat materials for which a lower temperature is needed (polymers, chemical synthesis, cooking). Later in 1999, the melting of powder particles with microwave energy was reported. Microwaves are electromagnetic waves with wavelengths in the range of 1m to 1mm and frequencies of 300 MHz to 300 GHz. The microwave energy dissipates the electric energy into the heat throughout the volume. The applications like selective heating, remelting, and volumetric heating, with negligible microstructural and thermal defects, make microwave irradiation an efficient heating method

DOI: 10.1201/9781003248743-5

[1, 2]. It was observed that a significant increase in product quality in terms of microstructure and mechanical properties than the conventional sintered products. Later, several studies were conducted on the processing of different materials. The processing of ceramics with microwave energy was first reported in 1968. Since 1980, many research groups have started working on the processing of ceramic materials. The microwave absorption mainly depends on the structural characteristics of the material rather than the bulk temperature of multiphase multi-multicomponent systems. Some microwave heating characteristics include enhanced physical and mechanical properties, proficient processing of complex shapes, and various material processing modes (selective, uniform, and rapid heating). Tangible benefits of the microwave for materials processing include reducing processing costs, quality products, and being environmentally safe. With appropriate control of this technology, various technically important materials can be processed. Also, the molecular-level heating phenomenon leads to the unique quality of the products, which cannot be produced with other conventional processes. The dielectric response of various materials is an important aspect of understanding materials processing. The database for the dielectric properties of the materials with respect to frequency and temperature is essential to predict the heating characteristics of the particular material. Earlier studies mainly focused on microwave-assisted transport and heating characteristics [3–5]. Also, the heating characteristics of the material depend on skin depth. The critical temperature of the particular material varies based on skin depth. Among materials, ceramics reaches critical temperature very rapidly due to their higher skin depth and loss tangent. At the critical temperature, microwaves couple directly with materials without any external agent. However, excessive heating of the material can lead to thermal defects. For optimum heating and better mechanical properties, ceramic matrix composites were fabricated without any thermal runaways. Furthermore, susceptor heating was developed for indirect heating of the materials whose skin depth is low. The heating and cooling rates, the length of soaking time, and microwave absorption can be controlled by changing the composition of the susceptor. The skin depth of the materials is noted for silicon carbide (1.93), graphite powder (1.34–2.09), carbon black (5.75), and activated carbon (0.7–3.43). In recent years, microwave energy is applied in various fields such as chemistry, drying, communication, diagnosis, and materials processing, among others.

5.1.2 Microwave Usage for Melting of Metallic Materials

The bulk metals of the penetration depth are typically in the range of 0.01–10 µm, owing to the significant reflection of microwaves. However, metallic powders of d/D_p ratio of about 2.4 particle size can be heated via coupling with microwaves [6, 7]. The role of the susceptor is crucial for material processing. The very idea of using susceptor in materials processing is to increase the absorption of microwaves. The ability to absorb microwaves in microwave heating includes dielectric loss, grain size, electrical conductivity, magnetic properties, frequency, porosity, and so on. As microwave heating promotes inverse heating, it generates heat internally rather than from external sources. It can also note that microwave heating happens mainly because of energy conversion rather than energy transfer. The microwave power and susceptor materials are significant in order to ensure the optimum melting of target materials. The microwave absorbing characteristics of the material would vary concerning the electric and magnetic field strengths throughout the material thickness. Based on the microwave absorption characteristics, materials can be divided into four categories such as transparent, absorber, opaque, and mixed absorbers. Among those, mixed absorber materials such as silicon carbide and charcoal are widely used as susceptor materials for processing metal matrix composite (MMC)-, polymer matrix composite (PMC)-, and ceramic matric composite (CMC)-based composite systems. Many studies have been focused on microwave cladding [8], microwave joining [9, 10], microwave sintering [11]. Also, many studies were reported on the processing of MMCs [12–16]. It was reported earlier that the addition of reinforcements of having high dielectric properties can be added to improve the heating rate and their mechanical properties.

5.1.3 Role of Microwave Route Technology for Processing of Materials

The microwave melting can be used for sintering, casting, cladding, and joining, among others. Despite the competitive welding techniques and thermal spray coatings, microwave heating remains indispensable in several applications due to the possibility of obtaining better quality products and rapid processing time and providing excellent microstructure. Also, it is noted that microwave heating decreases 3–5 times CO_2 emission than conventional heating. Compared with conventional heating methods, the microwave heating method reduces the energy consumption by 60–80%, which is attributed to the interaction of electric and magnetic fields that result in heating with dielectric loss, joule loss, and magnetic loss. The comparison study of melting aluminum with both conventional and microwave heating is shown in Figure 5.1.

Comparatively, the processing time of aluminum melting with microwave energy was reduced by 2.5 approximately. The materials processing with conventional methods has many drawbacks such as non-uniform heating, longer processing time, higher power consumption, and so on [18]. Generally, methods used for materials processing can be divided into two categories: contact

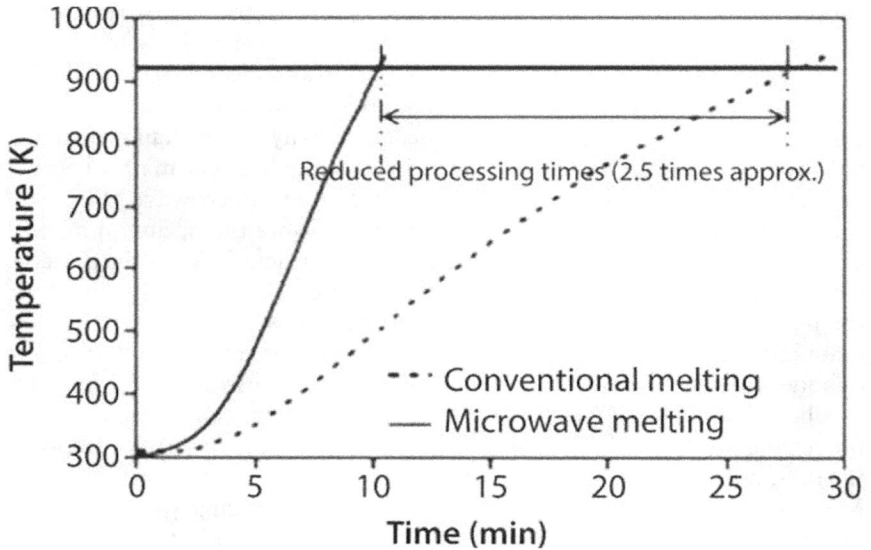

Figure 5.1 Comparison of microwave melting with the conventional melting of aluminum [17].

and noncontact type methods. The contact methods are traditional, using electric, resistant, and fuel heating sources, whereas the noncontact type utilizes induction, radiofrequency, and microwave heating. In comparison with the conventional methods, microwave processing is characterized as follows:

- Digital monitoring and control of system process parameters and material processing are possible.
- As the process is a noncontact type of materials processing, no thermal distortion and thermal defects will occur.
- The melting of materials will occur due to the propagation of electromagnetic waves.
- Due to the molecular-level heating occurring from inward to outward rather than outward to inward, the processed parts show good microstructure.
- Comparatively, a minimum shielding medium is needed while working with highly reactive materials like titanium, niobium, and molybdenum, among others.
- There is more reproducibility in the properties of materials processed by microwave energy compared to the conventional processing of materials.
- Joining or cladding of dissimilar metals that other conventional fabricating methods.

The efficiency of the microwave process mainly depends on the physical and chemical properties of the material and system parameters. Owing to its capability of microwave energy, manufacturing processes like brazing, cladding, sintering, and microwave-assisted machining can be done, which has drawn significant attention recently in industry and academia. For example, in conventional welding processes like tungsten inert gas (TIG), metal inert gas (MIG), electron bean welding (EBW), and laser beam welding (LBW), heat propagation directly affects the workpiece because of high heat exposure, propagating thermal cracks, and causes a change in microstructure, among others. The processes of laser beam and electron beam welding require a vacuum, which is difficult to maintain and increases the process cost. In the microwave joining process, selective heating takes place where only the parent metal is exposed to radiation. Similarly, in conventional sintering, the entire chamber heats up due to the transfer of energy from the surface to the core of the material. In microwave sintering, heating starts from the inner core of the component and reaches the surface. The physical vapor deposition (PVD) and chemical vapor deposition (CVD) processes require higher processing time and are only suitable for small-scale production. Also, the cost of the products is very high, which limits its applications in the manufacturing industry. In conventional casting techniques, manual selection of process parameters is necessary while producing an optimal processed product. In microwave casting, uniform melting would significantly reduce time due to the rapid energy transfer between matrix and reinforcement. Compared with conventional casting, microwave casting takes only one-third of the energy. The drilling processes operated by energies, such as laser, electron, and ion beams, require larger sophisticated processing equipment, extending processing and maintenance costs. However, microwave drilling works by the selective heating principle for localized melting of the target material. To overcome the limitations of conventional welding, sintering, cladding, and casting, microwave energy can effectively be applied to various manufacturing processes. The heat generation in microwave processing mainly depends on the magnetic and dielectric properties of the materials. The commercialization of microwave processing is needed to extend its applications. The microwave processing of the materials is gained attraction recently due to the enhancements in reaction and diffusion kinetics, comparatively low cycle time, improved energy savings, and so on. The processing of various materials with conventional processing methods leads to various limitations such as higher processing times, poor material quality, hazardous environment, and so on. However, limitations such as the reflectivity of metals in bulk material processing can be overcome by hybrid heating technology. Initially, microwaves are used for communication systems, including radar and broadcasting. Due to its rapid heating mechanism, microwave heating was later used for materials processing [19, 20]. The microwave effect usually happens while processing the ceramic materials because of their rising loss tangents, but the lower thermal conductivity of the ceramics causes the formation of hot spots. At these hot spots, irreversible damage happens because of excessive microwave heating, which can also be termed "thermal runaways." The metallic matrix can be added [21–23] to reduce the microwave effect and increase the thermal conductivity of the materials. In metal matrix and ceramic matrix composites, usually, one phase reflects microwaves where other phases absorb microwaves, enhancing the heating mechanism. In contrast to conventional methods, microwave processed components feature very good microstructure without any porosity and thermal defects. Also, these processed components are very close to final contours, which can ultimately reduce the cost of finishing. It has been studied that the melting characteristics of aluminum with microwave energy and conventional methods and reported that microwave processing reduced the power savings and time by a factor of 4 and 7, respectively [24]. Similarly, the comparative study of the sintering of W-Fe-Ni alloy with both conventional and microwave processing has been reported. It was evident from the results that the microwave processing time was reduced by 4 times than the conventional sintering process [25].

5.2 Microwave Cladding of Metallic Materials

In the present scenario, surface engineering is an important aspect, which offers superior bulk properties on the surface. Numerous studies have been conducted for surface modification of metallic surfaces, such as welding based methods Laser [26], TIG [27], plasma transferred arc (PTA) [28], thermal spraying (plasma, high velocity oxy fuel (HVOF), high velocity air fuel (HVAF), flame spraying) [29, 30], ion implementation [31], vapor deposition (PVD/CVD) [32, 33]. Among those, arc-assisted processes such as TIG, laser, and PTA, the cladding process provides dense microstructure and strong metallurgical bonding with the substrate because of higher fusion temperatures in the process. The thermal spray method can produce components with reproducibility and controllability, but low bond strength and higher porosity limit applications. The limitations such as porosity and low bond strength can be overcome by applying arc-assisted surface engineering methods. The processes such as laser, TIG, and PTA cladding techniques offer good components with better mechanical properties, bond strength, and low porosity. However, these processes consume high power, produce harmful gases, poor controllability, and induce defects like thermal distortion and porosity. The combination effect of uniform heating and rapid solidification profoundly influences the microstructures of the deposited materials. Also, rapid solidification extends the solid solubility limit and eliminates elemental partitioning, thereby inducing the formation of metastable phases. The microwave hybrid heating arrangement is applied for processing bulk materials that cannot be processed by direct microwave heating [34, 35]. The susceptor heating is the

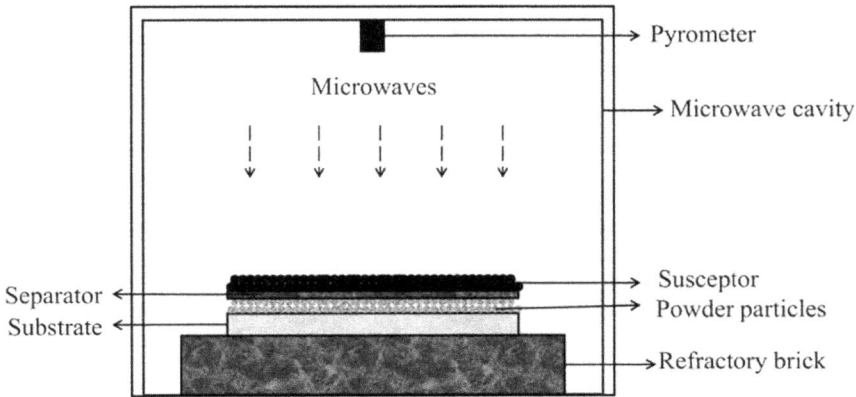

Figure 5.2 Schematic of the microwave hybrid heating setup.

main arrangement for the microwave hybrid heating mechanism. Initially, the target material absorbs heat from the susceptor material through conduction and convection. The target material gets heated up until the critical temperature. After reaching critical temperature, the target material starts absorbing microwaves directly results in rapid heating. The schematic representation of microwave hybrid heating is shown in Figure 5.2.

Although the skin depth of metallic materials is low, most the metallic powders can be processed using microwave energy with the change in particle size. Cladding is widely applied in mining, petrochemical, and gas industries to extend the service life of the components. The components in these industries experience severe corrosion, impact, adhesive, and abrasive wear. Usually, cladding materials deposited with various cladding techniques such as TIG, laser, MIG, and PTA induce residual stresses due to the variation of thermal gradient between substrate and coating. The interaction of metallic powders with microwaves was first reported while processing ceramics. The addition of the reactive metal powder aluminum was improved the heating rate. The higher d/D_p ratio of the particles can absorb microwaves, whereas small particles usually absorb microwaves by conduction- and convection-mode heat transfer [36]. Material systems such as cobalt-based, ferrous-based, and nickel-based are used for many applications. But nickel-based alloys are widely applied in the industries due to their superior material characteristics against corrosion, oxidation, adhesive, and abrasive wear.

5.2.1 *Effect of Process Parameters on the Clad Quality*

The significant parameters involved in microwave processing are electric and magnetic properties, load, applicator, and mechanism involved in the operation. These parameters are further influenced by factors like morphological and particle size of the powder, applicator type, and so on. The factors affecting the clad-layer microstructure can be classified into two categories, namely, intrinsic and extrinsic conditions. Intrinsic conditions are material-related, such as composition, volume fraction, particle size, and so on. The extrinsic conditions are mainly referred to as power, exposure time, and the like. In the cladding process, material properties such as reaction kinetics of the material and chemical affinity play a vital role that would eventually change the coating characteristics. Empirical models can be used to predict the optimum process parameters for better material characteristics. Usually, conventional processes possess thermal defects in the material. But microwave-processed components show no signs of thermal defects. The thermal stresses induced in the cladding can be expressed as follows:

$$\sigma_{th} = \frac{E\Delta_a\Delta_T}{1-v} \, ,$$

where E and v are the elastic modulus and Poisson's ratio of the clad material, respectively; D_α indicates the difference of thermal expansion between substrate and cladding; and D_T indicates the temperature difference between substrate and cladding. The previous studies revealed that optimum process conditions could significantly affect the microstructure and mechanical properties of the processed part. To achieve the desired properties optimization of process parameters, the design of suitable cladding materials is necessary. The earlier studies have also demonstrated that controlling the melting characteristics through careful selection of microwave power, susceptor material, exposure time, and materials compatibility would influence the microstructure of the processed part. To enhance the overall performance of the claddings, the proper design of the composite system in the fabrication attracts much interest in recent years. The supporting ceramic plates used in the microwave processing would alter the heating effects in the sample.

5.2.2 Selection of Materials

Materials with a good strength-to-weight ratio, higher stiffness, and wear resistance are needed for many industrial applications. The fabrication of these materials to satisfy industrial needs cannot be achieved by a single material. In the present era, composite materials are used in almost every area [37]. The higher impact toughness of the material is needed for most of the applications. The MMCs provide very good impact toughness that cannot be achieved by polymers. With the increase of metal resources on a large scale, in particular, the applications of alloys have expanded into the aerospace, medical, chemical, and automobile industries, owing to their extraordinary physical and mechanical properties. The properties of metallic materials include low ductility, good conductivity and heat transfer, and so on. With the increase of industries and consistent demand for engineering alloys, their applications have extended than any other structural materials. The functional and mechanical properties of the alloys are usually strongly influenced by the thermal effects and change of chemical compositions associated with the applied fabrication. In contrast, decreasing the inhomogeneous distribution of the phases to enhance the microstructural properties is requisite for many applications. The selection of cladding materials is to be chosen based on the functional properties of the substrate surface. Multicomponent component coatings have attracted significant attention because of their efficient operating conditions at higher temperatures. Usually, the physical properties, such as modulus of elasticity (E), melting point (T_m), and thermal expansion coefficient (g), of both cladding and substrate material should be taken into consideration. Too much variation in the melting point makes it difficult to produce the metallurgical bonding between them [38]. Nowadays, composite materials that combine ceramics and metal or alloys are the most popular cladding materials. The structure property of the materials is an important concern for industrial needs. Previous studies are mainly confined to polymeric, ceramics, and inorganic materials. It is believed that all metallic materials reflect microwaves. But recently, many studies have been conducted on metallic materials in powder form and observed that powder particles of lower particle size absorb microwaves effectively [39]. Among material systems, nickel-based alloys have promising applications due to their exceptional properties like high bond strength, low melting point, self-fluxing, better oxidation and corrosion resistance, adhesive and abrasive wear behavior, and very good corrosion behavior [40, 41]. Nickel-based alloys are widely applied in roller mills, turbines, extruders, and piston rods due to their excellent tribological properties at higher temperatures [42]. Porosity is mainly related to the solubility of hydrogen in the parent material, which is the function of temperature. To minimize porosity and improve the microstructure of the microwave heated part, the optimum selection of system parameters and susceptor material is essential. Nickelbased alloys have promising applications due to their exceptional properties like high bond strength, low melting point, self-fluxing, better oxidation and corrosion resistance, adhesive and abrasive wear behavior, and very good corrosion behavior [41]. Due to the lower ductility and other functional properties, ceramic MMCs lead to cracking and induce several thermal defects. To enhance the surface properties, the addition of solid lubricant additives in the composite system is vital. The usage of lubricant additives in the microwave processing of materials is very limited. The demand for self-lubricating materials is increasing due to their exceptional properties at high-temperature applications. The addition of these liquid lubricants adds complexity, cost, and weight to the system. Also, liquid lubricants break down at higher temperatures, limiting their applications and performance under certain conditions. The wastage of lubricant oil and other associated problems degrade its environmental regulations. The elimination of liquid lubricants with solid lubricants increases the toughness and adhesive resistance and is helpful in minimizing the coefficient of friction. The thin film formed at the contact surface modifies the solid–solid interaction and avoids asperity welding.

5.3 Microstructural and Mechanical Characteristics of the Claddings

In recent years, many investigations have been carried out on the spatial distribution microwave processing of materials [43]. Many researchers have studied theoretical studies based on the thermal transport of microwave phenomena. The heating rate and the degree of uniformity are significant features in the process. Also, the developed microwave clad showed no signs of thermal distortion, porosity, and solidification cracks due to the volumetric heating.

Gupta et al. developed WC10Co2Ni cladding through microwave hybrid heating. It was observed that the developed clads were metallurgically bonded with the stainless-steel substrate due to the partial dilution of a thin layer of the substrate. It was evident that the processed clads had no cracks, porosity, and solidification cracks. The microstructure and microhardness of the developed clad are shown in Figure 5.3. Due to the uniform distribution of hard phases in the clad, microhardness values of the clad (1064 ± 99 Hv) have minimal deviation throughout the clad depth. Also, the developed clad showed a very high resistance to sliding wear [44].

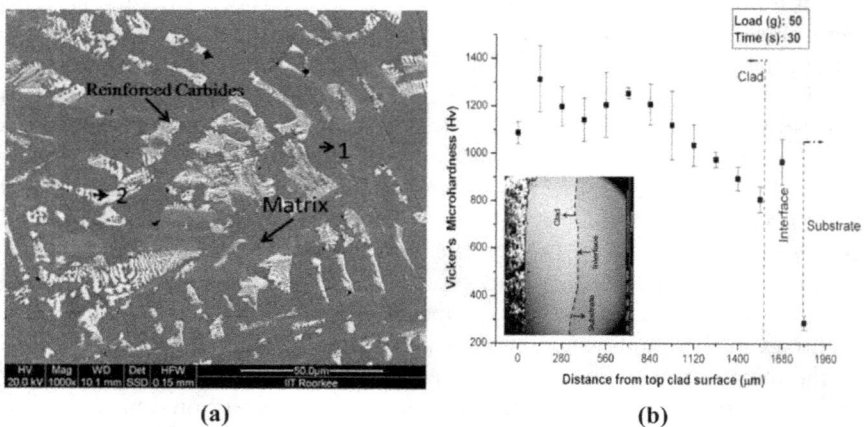

Figure 5.3 (a) Cross-sectional image of the clad showing skeleton structure; (b) microhardness values distribution throughout the clad cross section [44].

Similarly, Nair et al. developed high entropy alloy microwave treated claddings by varying Al fraction ($Al_xCoCrFeNi$) (x = 0.1 to 3). It can be evident from the microstructure that the clad was mainly composed of cellular structure and intermetallic phases, which were segregated and formed in the intercellular region (Figure 5.4). The interface was observed to be free of cracks and minimum porosity (<1%). Few secondary phases were also observed in Figure 5.4. The intercellular regions of equimolar and three molar Al were mainly composed of B2 and A2 phases. The average hardness of the clads was two- to threefold higher than the steel substrate. The results noted that the microwave-synthesized high-entropy claddings showed superior tribological properties for equimolar composition [45].

In another study, a nickel-based clad layer composed of EWAC + 20 % Cr23C6 was developed by microwave hybrid heating and reported that the clads had good microstructural and mechanical characteristics. The produced clads were free of porosity and other thermal defects. It was reported that the change of microstructure by cladding micro and nanostructured WC-12 Co materials. The clad produced with micrometric materials showed skeleton-type carbides in the metallic matrix. The cross-sectional scanning electron microscopy (SEM) images, both micrometric and nanometric clad, are shown in Figure 5.5. In comparison, nanostructured clad showed a uniform distribution of hard phases. A significant increase in hardness was observed with the decrease of particle size. The microhardness values of micrometric and nanometric clad were observed to be 1564 ± 50 HV and 1138 ± 90 HV. Due to the higher volume fraction of hard phases, a nanometric clad showed superior wear resistance. Compared to the micrometric clad, the nanometric clad exhibited 54% higher wear resistance. The grain

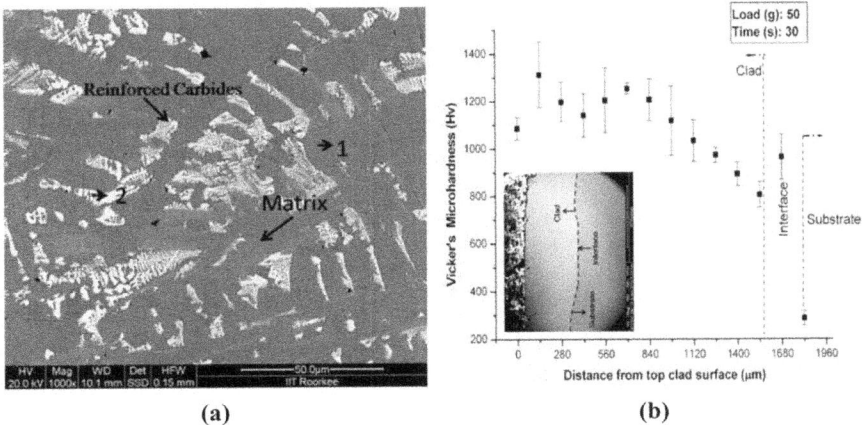

(a)   (b)

Figure 5.4 Optical and SEM images of (a, d, g) $Al_{0.1}CoCrFeNi$, (b, e, h) AlCoCrFeNi, and (c, f, i) $Al_3CoCrFeNi$ microwave-cladded high-entropy alloy claddings [45].

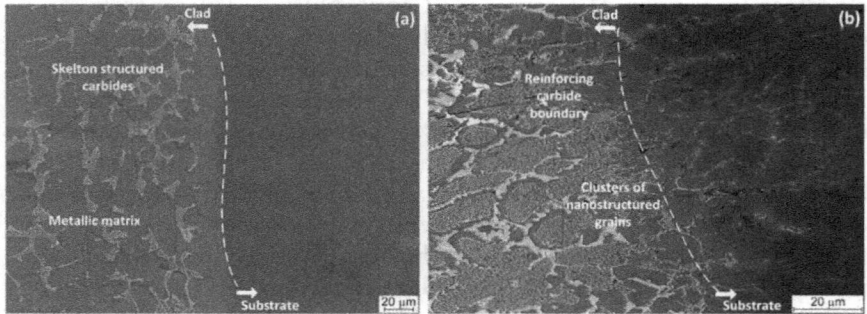

Figure 5.5 Cross-sectional SEM image of (a) WC-12Co micro-metric clad and (b) WC-12Co nano-metric clad [46].

pullout was noted in the micrometric clad due to the nonuniform distribution of carbides. However, the nanometric clad showed no three-body abrasion due to the excellent distribution of nano-carbides in the matrix [46].

Likewise, synthesized Ni-SiC composite claddings were developed using microwave irradiation. Different cladding systems were developed by varying weight fractions and particle size. The cross-sectional SEM images of the different claddings Ni, M5 (micro SiC wt%5), N5 (nano SiCwt% 5), B5 (nano SiC wt%2.5, micro SiCwt% 2.5), M10 (micro SiC wt%10), N10 (nano SiC wt%10), B10 (micro SiC wt5%, nano SiC wt%5) are shown in Figure 5.6. The microstructure of the claddings was mainly composed of a columnar structure with the formation of precipitated intermetallic compounds. The developed clads were free of porosity and metallurgically bonded with the substrate. Among all material systems, Ni-10 wt.% SiC showed superior cavitation erosion resistance, which was about 7–8 times higher than the stainless-steel substrate. The Ni–SiC bimodal materials showed the highest microhardness and fracture toughness. In microwave-cladded components, the splat boundaries and better adhesion were two significant effects that lead to resistance against cavitation erosion. It was observed that the microwave-synthesized Ni–SiC material systems were outperformed the thermal spray coatings [47].

In another study, Ni + 20 % Cr_7C_3 composite clads were developed on the CA6NM turbine steel using the microwave hybrid heating technique. The microstructural characteristics of the cladding material were studied by varying exposure times of 15 min, 25 min, 35 min, and 45 min. The clad processed at 35 min showed rod-like carbide structures. At 45 min exposure time, the microstructure of the clad was improved and showed fine dendritic type carbide structures. The cross-sectional SEM images of the clad

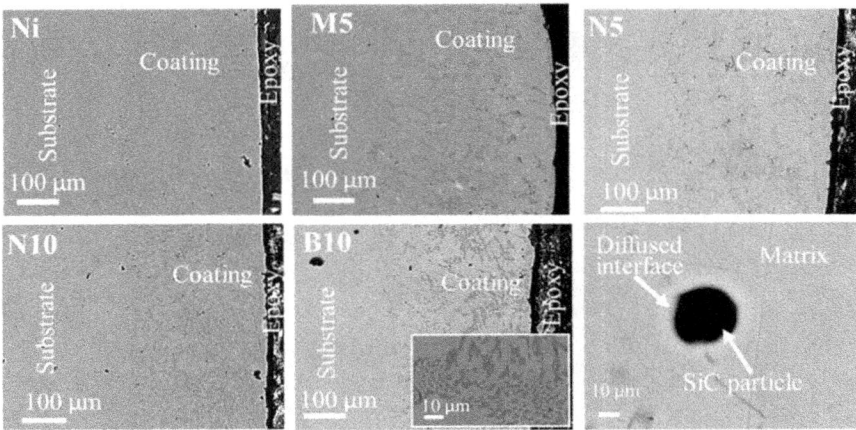

Figure 5.6 Cross-sectional SEM images of microwave-heated Ni-SiC composite claddings of different compositions [47].

processed at different exposure times are shown in Figure 5.7. From the results, the weight loss of the components cladded at 35 minutes and 45 minutes was reduced significantly by 29% and 45% than turbine steel substrate. The improved erosion resistance of sample 4 was attributed to the uniform distribution of dendritic phases.

The nano-hardness of individual phases for both matrix and reinforcements are shown in Figure 5.8a. It was observed that the nano-hardness of samples 3 and 4 was increased to two times that of sample 2. As indicated by XRD analysis, sample 2 was mainly constituted with Ni. In sample 3, hard phases like FeSi, Cr_3Si, and Ni_3C were detected. These hard phases were reported to be in the range of 9.3–11.77 GPa. Among four processed samples, sample 4 was mainly constituted of fine carbides.

These hard phases were found in the range of 19.3–21 Gpa, whereas nano-hardness values of coarse carbides in sample 3 were found in the range of 17–19.5 Gpa. It was noted that the mean nano-hardness value of the sample processed at 45 minutes was increased by 27% as compared to the sample processed at 35 minutes. The distributions of nano-hardness values of samples are shown in Figure 5.8b. The standard deviation of nano hardness of sample 4 was lower among the remaining samples, which might be due to the uniform distribution of a high-volume fraction of carbides.

The investigation of nickel- and cobalt-based clads were developed by various authors [49–51] in the recent past. The authors reported that the microwave clad developments and tribological behavior of the modified surface. There is very little work reported in slurry erosive wear studies of microwave

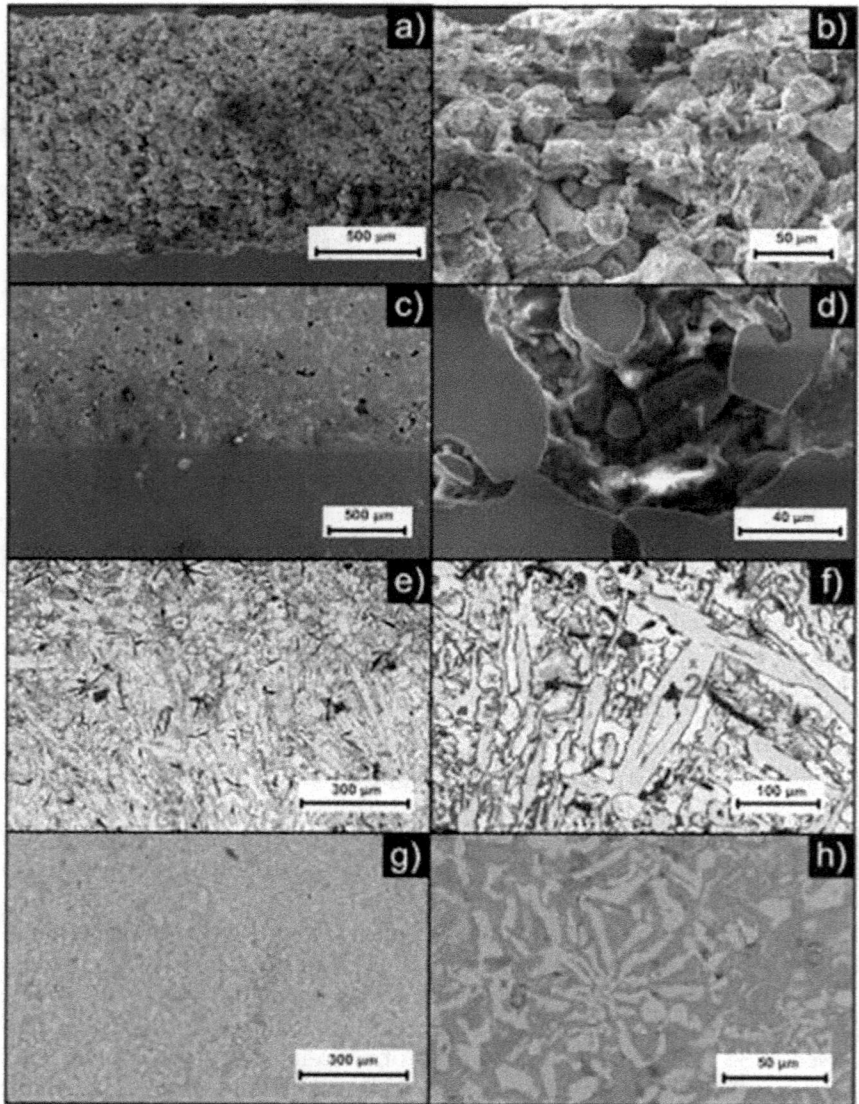

Figure 5.7 Cross-sectional SEM images of the samples produced at exposure time (a–b)
15 minutes—sample 1, (c–d) 25 minutes—sample 2, (e–f) 35 minutes—sample
3, and (g–h) 45 minutes—sample 4 [48].

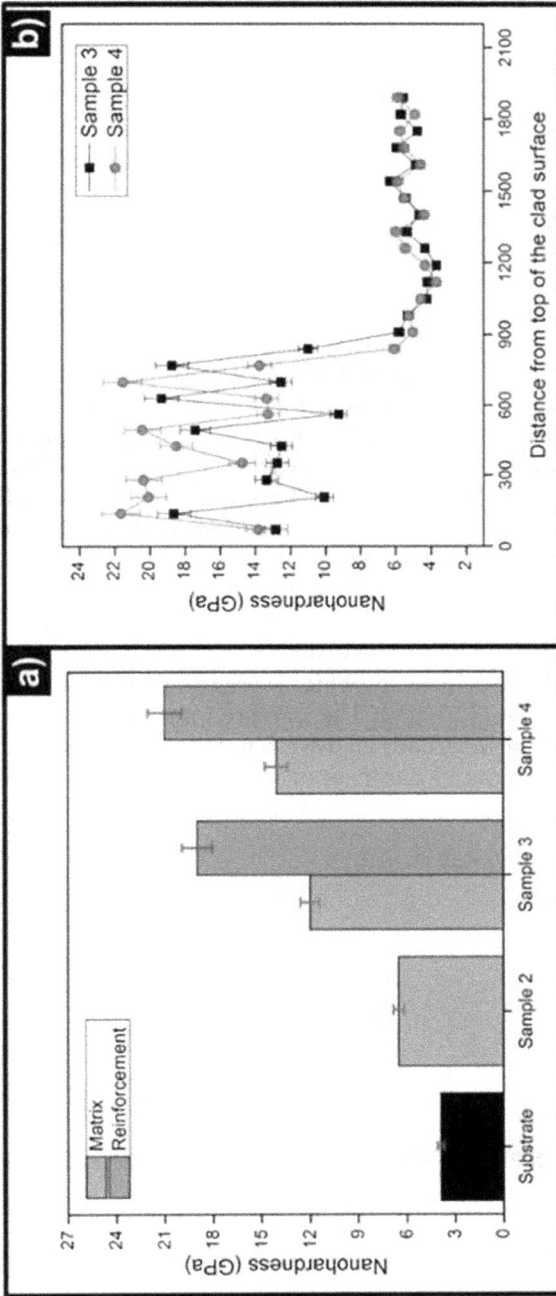

Figure 5.8 (a) Nano-hardness of various samples; (b) distribution of nano-hardness throughout the clad depth [48].

clads developed on austenitic and martensitic stainless steel and no reports on optimization of process parameters for slurry erosive wear studies through design of experiment (DOE) of the developed clads. The major contributions of the research are as follows. A nickel-based clad on SS-304 and cobalt-based clad on AISI-420 developed through domestic microwave irradiation. The developed clad surface proved to be excellent slurry erosive wear resistant, with an improved microhardness when compared to the substrate material.

The nickel alloy clads of NiCrSiB/WC, and NiCrSiB/WC/Ag/BaF$_2$were developed on Titanium 31 substrate by microwave hybrid heating technique. Both clads were metallurgically bonded with the substrate. No visible pores and cracks were observed in the clad due to the complete melting and subsequent solidification (Figures 5.9 and 5.10). The WC particles were well embedded in the nickel alloy matrix, and partial dissolution of the reinforcement phase was mainly responsible for the formation of hard phases in the clad. The presence of Cr in the clad further causes the dissolution of WC. The wavy interface observed in the clad indicates substrate dilution. The apparent needle-shaped particles in the clad were formed with W-Ni-Cr intermetallic phases. The microhardness profiles of both clads are shown in Figure 5.11. Due to the inverse thermal gradient phenomenon in the process, the minimal deviation was observed in the clad form clad surface to 800 μm. Whereas clads processed with other cladding techniques show more deviation in microhardness values due to the imbalance of thermal gradient. The average microhardness of the NiCrSiB/WC and NiCrSiB/WC/Ag/BaF$_2$ clads noted as 710 and 650 HV, respectively, which were comparatively higher than the titanium31 substrate (320 HV). The fracture toughness values of NiCrSiB/WC and NiCrSiB/WC/Ag/BaF$_2$ clad were noted as 5.7 and 6.35 Mpa.m$^{0.5}$, respectively. The enhanced hardness of the clad further improves the wear

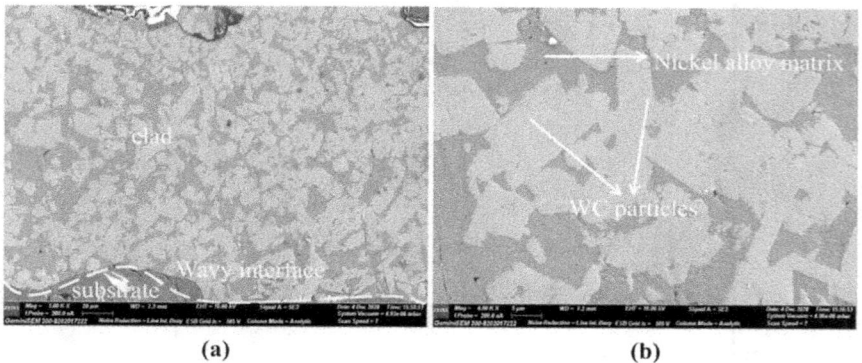

(a)   (b)

Figure 5.9 (a) A typical cross-sectional field emission scanning electron microscope image of NiCrSiB/WC microwave clad; (b) top region of the clad [52].

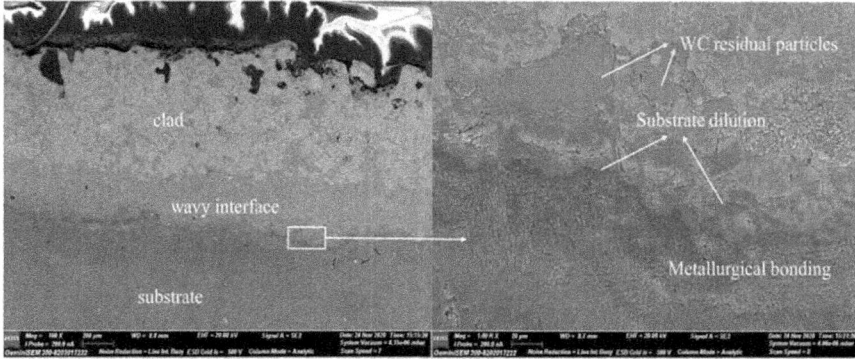

Figure 5.10 (a) The cross-sectional morphology of NiCrSiB/WC/Ag/BaF$_2$ clad; (b) near the interface [52].

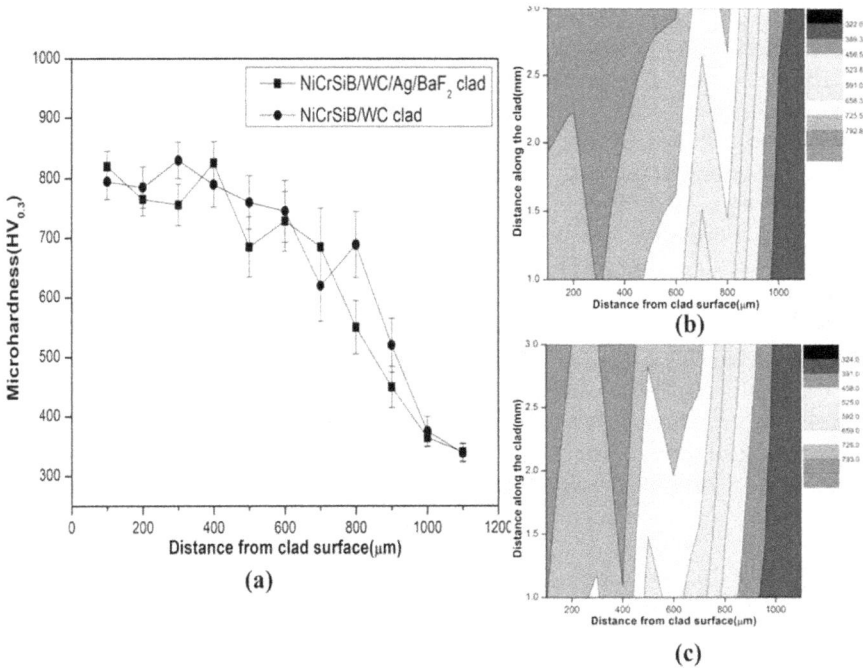

Figure 5.11 (a) Microhardness profile of NiCrSiB/WC and NiCrSiB/WC/Ag/BaF$_2$ clad, contour plot of microhardness across the cross section of (b) NiCrSiB/WC and (c) NiCrSiB/WC/Ag/BaF$_2$ [52].

resistance. The wear resistance of the clads improved by approximately 9 times than the titanium 31 substrate. To sum up, microwave processing enhanced the surface properties significantly and can be used as an effective method [52].

5.4 Future Outlook

The numerical methods, such as finite element analysis, analytical models, and first principle, can be used to predict defects and melting characteristics, explain mechanisms, and perform experiments effectively. The microwave processing using a multimode cavity possesses several problems due to its complexity. The finite element method can be used to understand the time-heating characteristics using software like COMSOL. Also, empirical methods and numerical methods involving field probing measurements can be applied for predicting the thermal behavior inside the cavity, which can be helpful in understanding the heating characteristics further. In the coming years, microwave energy could be applied for new materials processing and various fields. In the next few years, the commercialization of microwave energy for various applications will be accomplished. The physics behind the processing of materials with microwave energy was not explored well. The phenomenon of absorption behavior of different material systems with respect to micro-wave energy was studied. As materials processed with microwave energy are considered a highly efficient method, the microwave was coupled with different processes to increase efficiency in recent years. The processes such as microwave-assisted plasma, microwave plasma-assisted CVD were used for materials processing. The integrity of the laser process with a microwave can be developed in the coming years to extend applications in materials processing. The other conventional processes can be coupled with the micro-wave energy route to extend its industrial applications. The development of cladding material systems with the addition of solid lubricant additives can be used to enhance the operating temperatures in tribological applications. Large-scale microwave systems can be developed by analyzing the design, materials, and interaction of microwaves with materials at a higher tempera-ture. To develop these systems, several parameters should be considered, like microwave generator, waveguide, and type of cavity.

5.5 Summary

The design of a proper composite system while considering its microwave melting characteristics may effectively overcome the limitations of conven-tional methods. Microwave hybrid heating induces excellent properties that should be widely applied to replace conventional methods, which can be used to further industrial scale progress. It can be concluded that the overall state of research on microwave heating in the welding and surface engineering–related areas is well developed and has the great potential to extend its application in industries.

References

[1] El Khaled D, Novas N, Gazquez JA, et al. Microwave dielectric heating: Applications on metals processing. *Renew Sustain Energy Rev* [Internet]. 2018;82:2880–2892. Available from: https://doi.org/10.1016/j.rser.2017.10.043.

[2] Qiao X, Xie X. The effect of electric field intensification at interparticle contacts in microwave sintering. *Sci Rep.* 2016;6:1–7.

[3] Massoudi H, Durney CH, Barber PW, et al. Electromagnetic absorption in multilayered cylindrical models of man. *IEEE Trans Microw Theory Tech.* 1979;27: 825–830.

[4] Weil CM. Absorption characteristics of multilayered sphere models exposed to UHF/microwave radiation. *IEEE Trans Biomed Eng.* 1975;BME-22:468–476.

[5] Ayappa KG, Davis HT, Crapiste G, et al. Microwave heating: an evaluation of power formulations. *Chem Eng Sci* [Internet]. 1991;46:1005–1016. Available from: www.sciencedirect.com/science/article/pii/000925099185093D.

[6] Sun J, Wang W, Yue Q. Review on microwave-matter interaction fundamentals and efficient microwave-associated heating strategies. *Materials (Basel)* [Internet]. 2016;9. Available from: www.mdpi.com/1996-1944/9/4/231.

[7] Ertugrul O, Park H-S, Onel K, et al. Effect of particle size and heating rate in microwave sintering of 316L stainless steel. *Powder Technol* [Internet]. 2014;253:703–709. Available from: www.sciencedirect.com/science/article/pii/S0032591013008085.

[8] Gupta D, Sharma AK. Microwave cladding: A new approach in surface engineering. *J Manuf Process* [Internet]. 2014;16:176–182. Available from: http://dx.doi.org/10.1016/j.jmapro.2014.01.001.

[9] Bansal A, Sharma AK, Kumar P, et al. Characterization of bulk stainless steel joints developed through microwave hybrid heating. *Mater Charact* [Internet]. 2014;91:34–41. Available from: www.sciencedirect.com/science/article/pii/S104458031400059X.

[10] Bansal A, Sharma AK, Kumar P, et al. Investigation on microstructure and mechanical properties of the dissimilar weld between mild steel and stainless steel-316 formed using microwave energy. *Proc Inst Mech Eng Part B J Eng Manuf* [Internet]. 2014;230:439–448. Available from: https://doi.org/10.1177/0954405414558694.

[11] Eugene WWL, Gupta M. Characteristics of aluminum and magnesium based nanocomposites processed using hybrid microwave sintering. *J Microw Power Electromagn Energy a Publ Int Microw Power Inst.* 2010;44:14–27.

[12] Tun K, Gupta M. Improving mechanical properties of magnesium using nano-yttria reinforcement and microwave assisted powder metallurgy method. *Compos Sci Technol.* 2007;67:2657–2664.

[13] Mula S, Sahani P, Pratihar S, et al. Mechanical properties and electrical conductivity of Cu-Cr and Cu-Cr-4% SiC nanocomposites for thermo-electric applications. *Mater Sci Eng A-structural Mater Prop Microstruct Process—Mater Sci Eng A-Struct Mater.* 2011;528:4348–4356.

[14] Bescher EP, Sarkar U, MacKenzie JD. Microwave processing of aluminum-silicon carbide cermets. *MRS Online Proc Libr* [Internet]. 1992;269:371–378. Available from: https://doi.org/10.1557/PROC-269-371.

[15] Cheng J, Roy R, Agrawal D. Experimental proof of major role of magnetic field losses in microwave heating of metal and metallic composites. *J Mater Sci Lett.* 2001;20:1561–1563.

[16] Lorenson CP, Ball MD, Herzig R, et al. The microwave heating behaviour of metallic-insulator composite systes. *MRS Proc* [Internet]. 2011/02/28. 1990;189:279. Available from: www.cambridge.org/core/article/microwave-heating-behaviour-of-metallicinsulator-composite-systes/B1A56FF3672884C82677B9 6E19FF537F.

[17] Panda SS, Singh V, Upadhyaya A, et al. Sintering response of austenitic (316L) and ferritic (434L) stainless steel consolidated in conventional and microwave furnaces. *Scr Mater.* 2006;54:2179–2183.

[18] Roy R, Agrawal D, Cheng J, et al. Full sintering of powdered-metal bodies in a microwave field. *Nature* [Internet]. 1999;399:668–670. Available from: https://doi.org/10.1038/21390.

[19] Rao KJ, Vaidhyanathan B, Ganguli M, et al. Synthesis of inorganic solids using microwaves. *Chem Mater* [Internet]. 1999;11:882–895. Available from: https://doi.org/10.1021/cm9803859.

[20] Jones DA, Lelyveld TP, Mavrofidis SD, et al. Microwave heating applications in environmental engineering: A review. *Resour Conserv Recycl* [Internet]. 2002;34:75–90. Available from: www.sciencedirect.com/science/article/pii/S092134490 100088X.

[21] Veronesi P, Leonelli C, Pellacani GC, et al. Unique microstructure of glass-metal composites obtained by microwave assisted heat-treatments. *J Therm Anal Calorim.* 2003;72:1141–1149.

[22] Menéndez JA, Arenillas A, Fidalgo B, et al. Microwave heating processes involving carbon materials. *Fuel Process Technol* [Internet]. 2010;91:1–8. Available from: http://dx.doi.org/10.1016/j.fuproc.2009.08.021.

[23] Padmavathi C, Panda S, Agarwal D, et al. Effect of microstructural characteristics on the corrosion behavior of microwave sintered stainless steel. *Mater Sci Technol Process.* 2006;517–528.

[24] Chandrasekaran S, Basak T, Ramanathan S. Experimental and theoretical investigation on microwave melting of metals. *J Mater Process Technol* [Internet]. 2011;211:482–487. Available from: http://dx.doi.org/10.1016/j.jmatprotec.2010.11.001.

[25] Upadhyaya A, Tiwari SK, Mishra P. Microwave sintering of W-Ni-Fe alloy. *Scr Mater* [Internet]. 2007;56:5–8. Available from: www.sciencedirect.com/science/article/pii/S1359646206006798.

[26] Weng Z, Wang A, Wu X, et al. Wear resistance of diode laser-clad Ni/WC composite coatings at different temperatures. *Surf Coatings Technol* [Internet]. 2016;304:283–292. Available from: http://dx.doi.org/10.1016/j.surfcoat.2016.06.081.

[27] Gudala S, Ramesh MR, Nallathambi SS. Evolution of microstructure and high-temperature tribological performance of self-lubricating nickel-based composite tungsten inert gas coatings. *J Mater Eng Perform* [Internet]. 2021;30:8080–8094. Available from: https://doi.org/10.1007/s11665-021-06008-4.

[28] Zhou Y xin, Zhang J, Xing Z guo, et al. Microstructure and properties of NiCrBSi coating by plasma cladding on gray cast iron. *Surf Coatings Technol* [Internet]. 2019;361:270–279. Available from: https://doi.org/10.1016/j.surfcoat.2018. 12.055.

[29] Ramesh MR, Prakash S, Nath SK, et al. Evaluation of thermocyclic oxidation behavior of HVOF-sprayed NiCrFeSiB coatings on boiler tube steels. *J Therm Spray Technol*. 2011;20:992–1000.

[30] Navas C, Colaço R, de Damborenea J, et al. Abrasive wear behaviour of laser clad and flame sprayed-melted NiCrBSi coatings. *Surf Coat Technol*. 2006;200:6854–6862.

[31] Baumann H, Bethge K, Bilger G, et al. Thin hydroxyapatite surface layers on titanium produced by ion implantation. *Nucl Instruments Methods Phys Res Sect B Beam Interact with Mater Atoms* [Internet]. 2002;196:286–292. Available from: www.sciencedirect.com/science/article/pii/S0168583X020 12983.

[32] Costa MYP, Venditti MLR, Cioffi MOH, et al. Fatigue behavior of PVD coated Ti-6Al-4V alloy. *Int J Fatigue* [Internet]. 2011;33:759–765. Available from: www.sciencedirect.com/science/article/pii/S0142112310002719.

[33] Zhu Y, Wang W, Jia X, et al. Deposition of TiC film on titanium for abrasion resistant implant material by ion-enhanced triode plasma CVD. *Appl Surf Sci* [Internet]. 2012;262:156–158. Available from: www.sciencedirect.com/science/ article/pii/S0169433212005910.

[34] Srinath MS, Sharma A, Kumar P. Investigation on microstructural and mechanical properties of microwave processed dissimilar joints. *J Manuf Process*. 2011; 13:141–146.

[35] Srinath MS, Sharma AK, Kumar P. A novel route for joining of austenitic stainless steel (SS-316) using microwave energy. *Proc Inst Mech Eng Part B J Eng Manuf* [Internet]. 2011;225:1083–1091. Available from: https://doi.org/ 10.1177/2041297510393451.

[36] Zhou J, Xu W, You Z, et al. A new type of power energy for accelerating chemical reactions: The nature of a microwave-driving force for accelerating chemical reactions. *Sci Rep*. 2016;6:25149.

[37] Christensen RM. Mechanics of composite materials. Natl. SAMPE Symp. Exhib. 1984.

[38] Weng F, Chen C, Yu H. Research status of laser cladding on titanium and its alloys: A review. *Mater Des* [Internet]. 2014;58:412–425. Available from: http://dx.doi.org/10.1016/j.matdes.2014.01.077.

[39] Agrawal D. Microwave sintering of ceramics, composites metal powders [Internet]. Sinter. Adv. Mater. Fundam. Process. Woodhead Publishing Limited; 2010. Available from: http://dx.doi.org/10.1533/9781845699949.2.222.

[40] Rodríguez J, Martín A, Fernández R, et al. An experimental study of the wear performance of NiCrBSi thermal spray coatings. *Wear* [Internet]. 2003;255: 950–955. Available from: www.sciencedirect.com/science/article/pii/S0043164 803001625.

[41] Sharma SP, Dwivedi DK, Jain PK. Effect of La2O3 addition on the microstructure, hardness and abrasive wear behavior of flame sprayed Ni based coatings. *Wear* [Internet]. 2009;267:853–859. Available from: www.sciencedirect.com/ science/article/pii/S0043164809000180.

[42] Miguel JM, Guilemany JM, Vizcaino S. Tribological study of NiCrBSi coating obtained by different processes. *Tribol Int* [Internet]. 2003;36:181–187. Available from: www.sciencedirect.com/science/article/pii/S0301679X02001445.

[43] Ohlsson T. Dielectric properties and microwave processing. In: Singh RP, Medina AG, editors. *Food Prop Comput Eng Food Process Syst* [Internet]. Dordrecht: Springer Netherlands; 1989. pp. 73–92. Available from: https://doi. org/10.1007/978-94-009-2370-6_3.

[44] Gupta D, Sharma AK. Investigation on sliding wear performance of WC10 Co2Ni cladding developed through microwave irradiation. *Wear* [Internet]. 2011;271:1642–1650. Available from: www.sciencedirect.com/science/article/ pii/S004316481100158X.

[45] Nair RB, Arora HS, Boyana A V, et al. Tribological behavior of microwave synthesized high entropy alloy claddings. *Wear* [Internet]. 2019;436–437:203028. Available from: https://doi.org/10.1016/j.wear.2019.203028.

[46] Zafar S, Sharma AK. Dry sliding wear performance of nanostructured WC-12Co deposited through microwave cladding. *Tribol Int* [Internet]. 2015;91: 14–22. Available from: www.sciencedirect.com/science/article/pii/S0301679X 15002789.

[47] Babu A, Arora HS, Singh H, et al. Microwave synthesized composite claddings with enhanced cavitation erosion resistance. *Wear*. 2019;422–423.

[48] Singh B, Zafar S. Effect of microwave exposure time on microstructure and slurry erosion behavior of Ni + 20 % Cr 7 C 3 composite clads. *Wear* [Internet]. 2019;426–427:491–500. Available from: https://doi.org/10.1016/j.wear.2018. 12.016.

[49] Hebbale AM, Badiger R, Srinath MS, et al. An experimental investigation of microwave developed nickel-based clads for slurry erosion wear performance using taguchi approach. *Metallogr Microstruct Anal*. 2020;9.

[50] Hebbale AM, Srinath MS. Microstructural investigation of Ni based cladding developed on austenitic SS-304 through microwave irradiation. *J Mater Res Technol* [Internet]. 2016;5:293–301. Available from: www.sciencedirect.com/ science/article/pii/S2238785416000156.

[51] Hebbale AM, Srinath MS. Microstructural studies of cobalt based microwave clad developed on martensitic stainless steel (AISI-420). *Trans Indian Inst Met*. 2018;71:737–743.

[52] Suresh G, Ramesh MR, Srinath MS. Development of self: Lubricating nickel based composite clad using microwave heating in improving resistance to wear at elevated temperatures. *Met Mater Int* [Internet]. 2021. Available from: https://doi.org/10.1007/s12540-021-01078-4.

Chapter 6

Analysis of Mechanical Properties and Microstructural Characterization of Microwave Cladding on Stainless Steel

C. Durga Prasad, Mahantayya Mathapati, Hitesh Vasudev, and Lalit Thakur

Contents

6.1 Introduction

Surface modification is a technique of changing the characteristics of the surface of a material in order to make them suitable for the desired application [1–2]. Effective surface area is governed by morphology which is greater than the macroscopic geometrical area almost every time [3–6]. Because the surface energy of liquids facilitates smooth surfaces, solid materials are usually manufactured from liquids. Surface alteration could be attained in many ways. Some of them include processes like chemical vapor deposition (CVD), physical vapor deposition (PVD), nitriding, cyaniding, and cladding, among others [7–11]. To attain desired properties at surface-level cladding has proved to be the one surface alteration method [12–15]. The surface film governs most of the functional properties, apart from the morphology. Surface energy results when an attractive force is felt between the surface atoms [16–20]. Cladding minimizes spallation and downstream erosion. It can also minimize material/component cost [21]. The material degradation, such as wear, corrosion, and oxidation, in various engineering applications commonly occurs when the structural component subjected to different

DOI: 10.1201/9781003248743-6

environmental conditions [22]. To overcome the drawbacks of available surface modification techniques, microwave cladding can be used as a novel form of material processing that can serve with better prospects [23].

Microwave heating is a unique processing technique with respect to processing materials compared to traditional heating methods, such as muffle furnace, tubular furnace and several other furnace heating processes [24–25]. The heat transfer in traditional furnace systems takes place from outside to inside this result in non-uniform heating and sometimes material can undergo thermal distortion due to excessive heating. This could affect the mechanical and metallurgical properties of the material. In microwave energy, the heat transfer originates from the inside region to the outside region of the material. The heat distribution takes place at the molecular level; this leads to uniform heating throughout the material, and one can expect a reduction in thermal distortion of material [26–27]. This heating method can be controlled easily to achieve the essential properties of a material. The microwave heating method is economical compared to both conventional and advanced thermal processing systems [28–30].

The main intent of this work was to develop MoCoCrSi+Cr_3C_2 clads by microwave hybrid heating (MHH) technique using a microwave applicator of 2.45 GHz frequency and a power of 900 W in order to make metal harder and characterize the development of clads for microstructure and mechanical properties. The outcomes of the tested cladding are discussed in this chapter.

6.2 Materials and Method

The present research deals with the selection of cobalt-based powder, and chromium carbide (Cr_3C_2) as cladding materials and SS-316 as the target substrate material. MoCoCrSi was the base powder that was mixed with composite powder (Cr_3C_2) in the proportion of 70% to 30%. Prasad et al. [22, 23] studied the cladding of similar powder and given the detailed chemical composition. The powder material has more strength and resistance to wear and corrosion. Cr_3C_2 has high hardness, and it is wear-resistant material used in various applications. The cladding of MoCoCrSi+Cr_3C_2 on stainless steel (SS-316) was produced as a result of MHH. Cladding on a substrate was done in the domestic microwave apparatus as shown in Figure 6.1a. The schematic representation of MHH is shown in Figure 6.1b.

6.3 Microwave Heating Process

The MHH is the new and effective method of absorption whereby powders with a low absorption coefficient are melted using microwave energy. To absorb microwave energy by composite powder particles, susceptor materials are usually used. The susceptor materials like silicon carbide, charcoal, and so on can absorb all the microwave radiations impinged on it. Silicon carbide was used as a susceptor material for cladding. The susceptor absorbs microwave radiation and increases the rate of heat transfer. However, still there is a demand to enhance the cladding performance by using a susceptor that increases the rate of heat transfer as shown in Figure 6.1b.

The surface was washed with acetone to remove unwanted impurities from the substrate. MoCoCrSi+Cr_3C_2 mixture was made into a slurry by mixing it with an adhesive (Araldite) so that the powder stays in place without dripping. The slurry was smeared to the cleaned surface. To prevent the loss of heat and damage to the cavity, the substrate was enclosed with a masking material made of alumina.

The main components in the case of microwave cladding are shown in Figure 6.2a, and the development cladding is shown in Figure 6.2b.

After the clad powder was applied on the substrate a thin sheet of separator material (graphite sheet) was placed before placing the susceptor material on top of it; the susceptor soaks up the microwave radiations, converts

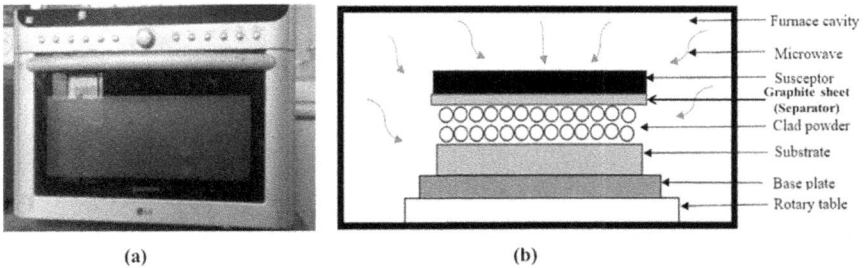

(a)   (b)

Figure 6.1 (a) Microwave applicator used; (b) schematic representation of the processing setup.

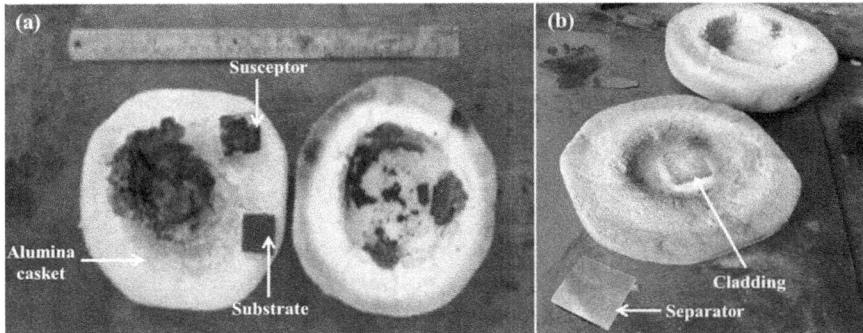

Figure 6.2 (a) Materials used in the cladding process; (b) development of cladding.

Table 6.1 Observation Table

Sl. No	Processing Time (min)	Observations
1	10	Sufficient heat is not supplied to melt particles.
2	25	Improper melting of clad powders occurs.
3	30	Powders are partially melted but poorly bonded.
4	35	Clad is partially developed but poorly bonded.
5	42	Clad is developed with partial melting of particles and good bonding with a substrate is observed.
6	45	Overheating takes place, which causes poor bonding, and the material gets deformed.

it to mechanical energy (heat), and transfers it to the clad material placed beneath it. To avoid adulteration, the separator was placed in between silicon carbide and $MoCoCrSi+Cr_3C_2$ powder, and to avoid heat loss, a refractory material (aluminum oxide/alumina) surrounds the setup. Alumina helps prevent the microwaves from escaping and to be confined within the setup as shown in Figure 6.2b. In the present work, several trials were carried out by varying the clad time from 35–42 min, whereas the power was kept constant at 900 W. It gives a clear picture that microwave exposure time is an important factor in the dilution of clad powder. The clad materials were exposed for 42 min to microwave radiation and were left to cool for 10 min between the successive experiments. The observations were noted during the development of cladding and listed in Table 6.1.

The cladding was developed after performing many trials. Initially, to optimize the processing of materials duration was varied and some observations with respect to time change were tabulated in Table 6.1. With an increase in processing time, the clad powders were exposed to elevated temperature for a longer duration; this resulted in the grains present at the substrate–cladding interface region migrating to attain a metallurgical bond. The exposure time was in direct relation to the bonding of the powder and the substrate. The homogeneity of the cladding surface was reliant on the frequency, heating method (i.e., normal or hybrid heating), and power of microwaves [3–7].

6.4 Microstructural Analysis

The cladding samples were taken for microstructural and microhardness studies. The mounted specimen was polished first by emery paper of 320 grit, then by using emery papers of grades 400, 600, 800, 1200, 1500, 2000, and finally with 1-micron diamond paste on a velvet cloth placed on the polishing machine, and the specimen was etched to enhance the appearance, prevent contamination, and remove oxidation to create a reflective surface and observe the crystal structure more effectively. After this, the etched specimen was observed using a metallurgical microscope to observe the grain structure and examined the presence of irregularities.

6.5 Mechanical Properties

The nano-indentation hardness (H) and the elastic modulus (E) of the sample cross section were measured as per ISO145774 and ASTM E-2546 standard using a nano-indentation instrument (UNHT, Switzerland CSM Instruments Co., Ltd.) with a diamond Berkovich indenter. The sample was polished thoroughly prior to the test. The polished surface was exposed to indentation, and a load of 30 g was applied to the sample for 5 s. Load-displacement data were obtained from a nano-indentation tester to analyze the hardness and elastic modulus of the cladding. The surface roughness of the cladding was more and confirmed the presence of unmelted/partially melted particles, resulting in uneven peaks on surface. Due to this reason, it is not possible to perform more indentation trails on the cladding specimen.

The scratch test was performed on cladding and substrate using a micro-scratch tester in ambient air, at the temperature of 23°C and humidity of 40% to measure the surface profile, friction coefficient, and critical load. The parameters, such as indenter (stylus)-type Rockwell diamond, stylus radius of 50 mm, a constant normal load of 4 N, a scratch length of 0.7 mm, and a stylus velocity of 1.4 mm/min, were used to produce three identical scratches on specimen.

Microhardness test was conducted across the surface of both the substrate and the cladding using Vickers microhardness tester. The test was carried out at three different sections across the cross-section of the clad sample. The indentations were made at 500-g load and a dwell time of 10 s.

6.6 Results and Discussion

6.6.1 Microstructure of Cladding

Clads were observed to have a better grain structure than the substrate (SS-316). Figure 6.3a shows the microstructure of the cladded substrate. The existence of minor signs of semi-melted particles trapped in the pores and voids can be seen in Figure 6.3a. The formation of irregular-sized and fine fragmented particles was identified in the microstructure depicted in Figure 6.3a. Also, in Figure 6.3b, a transverse view of the cladded substrate, it is noticed that the clad exhibits 550-μm thickness and bonded well with the substrate. The surface roughness of the cladding was measured and exhibits R_a 10.232 μm.

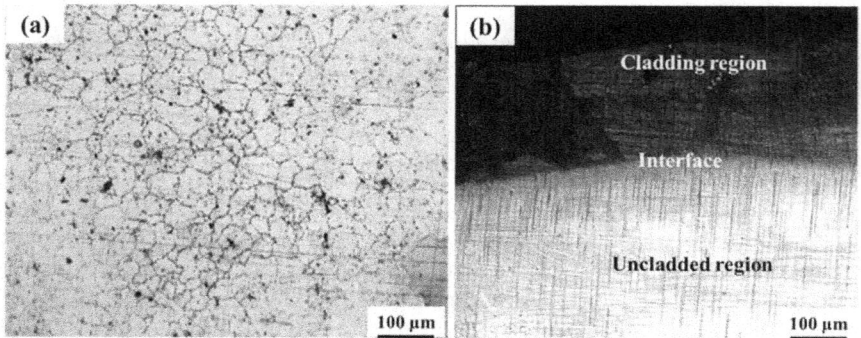

Figure 6.3 Microstructure of MoCoCrSi+Cr$_3$C$_2$: (a) Longitudinal view; (b) transverse view.

6.6.2 Nano-Indentation and Microhardness

Cladding elastic modulus and hardness were examined through load–displacement curve. The curve exhibits a loading and unloading trend for the four trails for the mechanical properties of cladding as shown in Figure 6.4a–c. The mechanical property results of cladding are tabulated in Table 6.2. The cladding reveals the presence of small voids, cracks, and rough surface, which causes the variation in results of hardness and modulus. The cladding exhibits better hardness and modulus.

Figure 6.5 depicts the surface scratch of the substrate and cladding samples. The scratch depth in substrate was increased due to relatively its soft nature as compared to the deposited cladding; hence, plastic deformation occurred when the load force reached a certain value. The strain hardening and the material stacking caused by the plastic deformation hindered the movement of the indenter after accumulating to a certain extent. Figure 6.5a displays that there are cracks perpendicular to the sliding direction in the scratch area, which were generated by the tensile stress of the indenter under dynamic contact [14, 19, 23–26]. Due to the increases in the applied load, the internal stress of the scratch groove and the crack density both increased, and wider cracks appeared. Figure 6.5b shows the scratch on the cladding sample surface. This scratch is shallow, gentle, and smooth without fragmentation and stratification, indicating that the cladding layer with a gradient structure has excellent adhesion to the substrate [18]. The lines in the grooves of the cladding propagated along the scratch direction.

The scratch depth of the cladding sample slowly increased with the applied load force increase its maximum depth was 18 µm without the bounce phenomenon. In contrast, the scratch depth of the substrate sample nonlinearly increased with the applied load force increase. The scratch

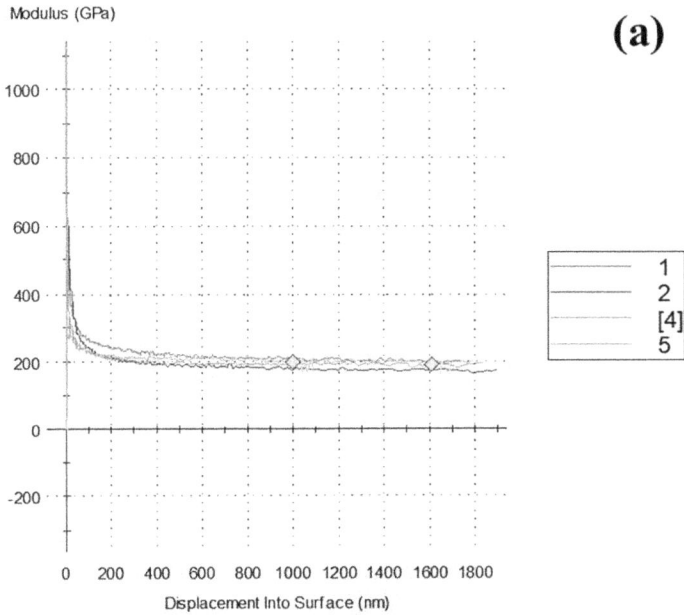

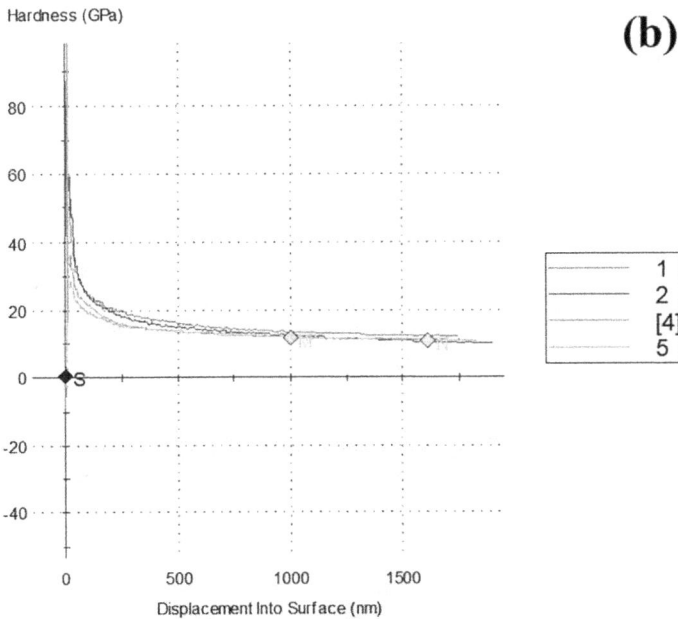

Figure 6.4 Cladding nano-indentation results: (a) Elastic modulus penetration plot; (b) hardness penetration plot; (c) load penetration plot during the indentation test.

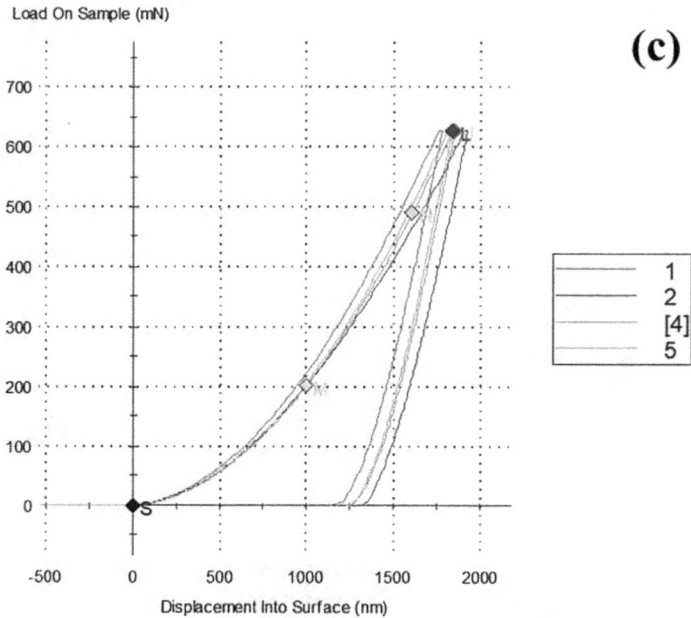

Figure 6.4 (Continued)

Table 6.2 Mechanical Properties of Cladding

Properties	Cladding	Standard Deviation
Hardness (GPa)	11.25	0.69
Elastic Modulus (GPa)	193.2	12.6

process has different energy dissipation mechanisms. The material stacking and the strain hardening caused by the plastic deformation will produce abnormal contact points between the indenter and the material, which will result in changes in the local depth and friction [10–14]. The maximum scratch depth of the substrate sample was 42 µm, which is 2.33 times that of the cladding sample. The friction coefficient of the cladding was 0.31, while that of the substrate was 0.76, which is 2.45 times that of the cladding. Therefore, compared with the substrate, the cladding has better wear resistance.

The cladded area was observed to have a greater microhardness value than that of the boundary and substrate area. The average microhardness of the substrate is found to be 285.66 Hv, while that of the MoCoCrSi+Cr_3C_2 clad

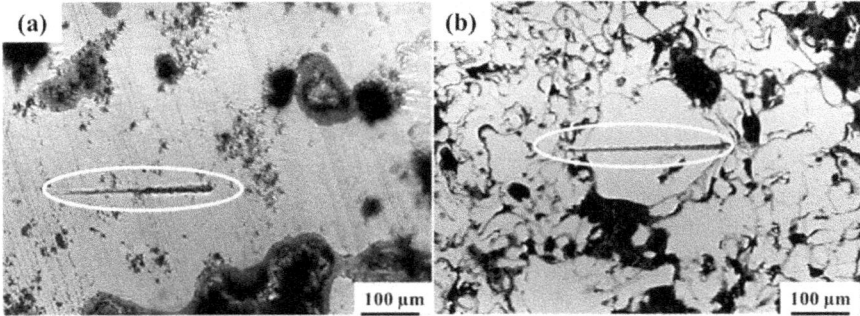

Figure 6.5 Micrograph of the sample scratch track: (a) Substrate; (b) cladding.

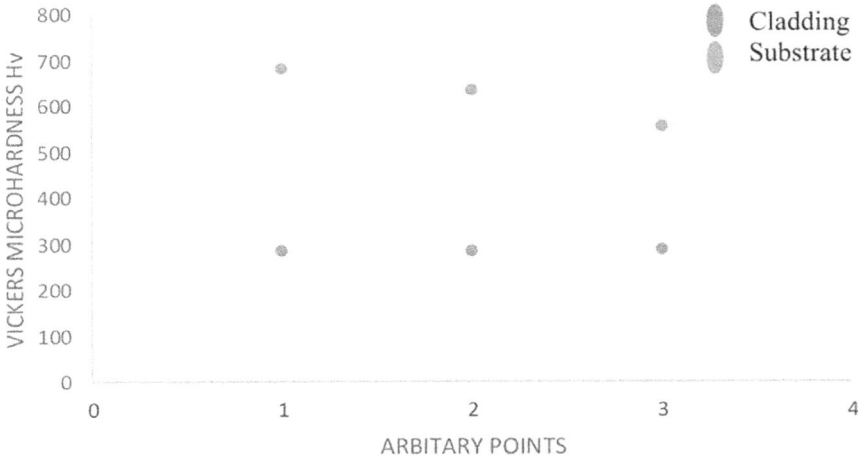

Figure 6.6 Vickers microhardness test report of cladded and substrate.

is 682.33 Hv. The test report shows a significant increase in the hardness of the sample. The microhardness plot is shown in Figure 6.6. On examination, it is found that the cladded grain size is smaller than the substrate, it depicts the better hardness of the clad formed. The fine-grain structure has better mechanical properties than the coarse-grain structure materials.

6.7 Conclusion

$MoCoCrSi+Cr_3C_2$ cladding on SS-316 substrate was developed by the MHH technique in order to characterize the clad with respect to mechanical and metallurgical properties. Conclusions were made from the present work is as follows:

- MoCoCrSi+Cr_3C_2 clad can be developed by the MHH technique.
- A microstructural study reveals defect-free grain structure on the surface of the developed clads and better interfacial bonding.
- Cladding reveals variation in elastic modulus and hardness properties due to presence of small voids and cracks.
- The clad sample exhibits better resistance to scratch by showing narrow and smooth groove as it confirms cladding has good wear resistance than the substrate.
- The hardness of the developed clad was found to be notably higher (~682.33 HV) than that of the uncladded substrate (285.6 HV).

References

[1] A. K. Sharma, S. Aravindan, R. Krishnamurthy, "Microwave glazing of alumina-titanium ceramic composite coatings", *Material Letter* 50 (2001), 295–301.

[2] Chun-Ming Lin, Shih-Hung Yen, Cherng-Yuh Su, "Measurement and optimization of atmospheric plasma sprayed CoMoCrSi coatings parameters on Ti-6Al-4V substrates affecting microstructural and properties using hybrid abductor induction mechanism", *Measurement* 94 (2016), 157–167.

[3] Thavaraya Naik, Mahantayya Mathapathi, C. Durga Prasad, H. S. Nithin, M. R. Ramesh, "Effect of laser post treatment on microstructural and sliding wear behavior of HVOF sprayed NiCrC and NiCrSi coatings", *Surface Review and Letters*. https://doi.org/10.1142/S0218625X2250007X.

[4] C. Durga Prasad, Sharnappa Joladarashi, M. R. Ramesh, Anunoy Sarkar, "High temperature gradient cobalt based clad developed using microwave hybrid heating", *American Institute of Physics* 2018 (1943), 020111. https://doi.org/10.1063/1.5029687.

[5] G. Madhusudana Reddy, C. Durga Prasad, Gagan Shetty, M. R. Ramesh, T. Nageswara Rao, Pradeep Patil, "High temperature oxidation studies of plasma sprayed NiCrAlY/TiO_2 & NiCrAlY /Cr_2O_3/YSZ cermet composite coatings on MDN-420 special steel alloy", *Metallography, Microstructure and Analysis*. https://doi.org/10.1007/s13632-021-00784-0.

[6] D. Gupta, A. K. Sharma, "Development and microstructural characterization of microwave cladding on austenitic steel", *Surface and Coatings Technology*. 205 (2011), 5147–5155.

[7] Madhu Sudana Reddy, C. Durga Prasad, Pradeep Patil, M. R. Ramesh, Nageswara Rao, "Hot corrosion behavior of plasma sprayed NiCrAlY/TiO_2 and NiCrAlY/Cr_2O_3/YSZ cermets coatings on alloy steel", *Surfaces and Interfaces* 22 (2021), 100810. https://doi.org/10.1016/j.surfin.2020.100810.

[8] C. Durga Prasad, Sharnappa Joladarashi, M. R. Ramesh, M. S. Srinath, "Microstructure and tribological resistance of flame sprayed CoMoCrSi/WC-CrC-Ni and CoMoCrSi/WC-12Co composite coatings remelted by microwave

hybrid heating", *Journal of Bio and Tribo-Corrosion* 6 (2020), 124. https://doi. org/10.1007/s40735-020-00421-3.

[9] C. Durga Prasad, Sharnappa Joladarashi, M. R. Ramesh, "Comparative investigation of HVOF and flame sprayed CoMoCrSi coating", *American Institute of Physics* 2247 (2020), 050004. https://doi.org/10.1063/5.0003883.

[10] E. T. Thostenson, T. W. Chou, "Microwave processing: Fundamentals and applications", *Composites A* 30 (1999), 1055–1071.

[11] C. Durga Prasad, Akhil Jerri, M. R. Ramesh, "Characterization and sliding wear behavior of iron based metallic coating deposited by HVOF process on low carbon steel substrate", *Journal of Bio and Tribo-Corrosion* 6 (2020), 69. https://doi.org/10.1007/s40735-020-00366-7.

[12] Ajit M. Hebbale, M. S. Srinath, "Taguchi analysis on erosive wear behavior of cobalt based microwave cladding on stainless steel AISI-420", *Measurement* 99 (2017), 98–107.

[13] Yang Li, Xiufang Cui, Guo Jin, Zhaobing Cai, Na Tana, Bingwen Lu, Zonghong Gao, "Interfacial bonding properties between cobalt-based plasma cladding layer and substrate under tensile conditions", *Materials and Design* 123 (2017), 54–63.

[14] A. K. Sharma, D. Gupta, "A method of cladding/coating of metallic and non-metallic powders on metallic substrates by microwave irradiation", Indian Patent 527/Del/2010.

[15] C. Durga Prasad, Sharnappa Joladarashi, M. R. Ramesh, M. S. Srinath, B. H. Channabasappa, "Comparison of high temperature wear behavior of microwave assisted HVOF sprayed CoMoCrSi-WC-CrC-Ni/WC-12Co composite coatings", *Silicon*, Springer (2020), 1–19. https://doi.org/10.1007/s12633-020-00398-1.

[16] Dheeraj Gupta, Prabhakar M. Bhovi, Apurbba Kumar Sharma, Sushanta Dutta, "Development and characterization of microwave composite cladding", *Journal of Manufacturing Processes* 14 (2012), 243–249.

[17] Apurbba Kumar Sharma, Dheeraj Gupta, "On microstructure and flexural strength of metal—ceramic composite cladding developed through microwave heating", *Applied Surface Science* 258 (2012), 5583–5592.

[18] T. Bell, "Surface engineering of Austenitic Stainless Steel", *Surface Engineering* 18 (6) (2002), 415–422.

[19] M. S. Srinath, A. K. Sharma, "Investigation on microstructural and mechanical processed dissimilar joints", *Journal of Manufacturing Processes* 13 (2011), 141–146.

[20] N. K. Bhoi, H. Singh, S. Pratap, P. K. Jain, "Microwave material processing: A clean, green, and sustainable approach", *Sustainable Engineering Products and Manufacturing Technologies* (2019), 6–7.

[21] K. P. Kumar, A. Mohanty, M. L. Lingappa, M. S. Srinath, S. K. Panigrahi, "Enhancement of surface properties of austenitic stainless steel by nickel-based alloy cladding developed using microwave energy technique", *Materials Chemistry and Physics* 256 (2020), 123657.

[22] C. Durga Prasad, Shashank Lingappa, Sharnappa Joladarashi, M. R. Ramesh, B. Sachin, "Characterization and sliding wear behavior of CoMoCrSi+Flyash composite cladding processed by microwave irradiation", *Materials Today Proceedings* 46 (2021), 2387–2391. https://doi.org/10.1016/j.matpr.2021.01.156.

[23] C. Durga Prasad, Sharnappa Joladarashi, M. R. Ramesh, M. S. Srinath, B. H. Channabasappa, "Effect of microwave heating on microstructure and elevated temperature adhesive wear behavior of HVOF deposited CoMoCrSi-Cr_3C_2 composite coating", *Surface and Coatings Technology* 374 (2019), 291–304. https://doi.org/10.1016/j.surfcoat.2019.05.056.

[24] Fei Weng, Huijun Yu, Chuanzhong Chen, Jianli Liu, Longjie Zhao, Jingjie Dai, Zhihuan Zhao, "Effect of process parameters on the microstructure evolution and wear property of the laser cladding coatings on Ti-6Al-4V alloy", *Journal of Alloys and Compounds* 692 (2017), 989–996.

[25] C. Durga Prasad, Sharnappa Joladarashi, M. R. Ramesh, M. S. Srinath, B. H. Channabasappa, "Development and sliding wear behavior of Co-Mo-Cr-Si cladding through microwave heating", *Silicon* 11 (2019), 2975–2986. https://doi.org/10.1007/s12633-019-0084-5.

[26] S. Zafar, A. K. Sharma, "Development and characterizations of WC—12Comicrowave clad", *Materials Characterization* 96 (2014), 241–248.

[27] C. Durga Prasad, Sharnappa Joladarashi, M. R. Ramesh, M. S. Srinath, B. H. Channabasappa, "Microstructure and tribological behavior of flame sprayed and microwave fused CoMoCrSi/CoMoCrSi-Cr_3C_2 coatings", *Materials Research Express* 6 (2019), 026512. https://doi.org/10.1088/2053-1591/aaebd9.

[28] M. Gupta, W. L. E. Wong, "Enhancing overall mechanical performance of metallic materials using two-directional microwave assisted rapid sintering", *Scripta Materialia* 52 (2005), 479–483.

[29] C. Durga Prasad, Sharnappa Joladarashi, M. R. Ramesh, M. S. Srinath, B. H. Channabasappa, "Influence of microwave hybrid heating on the sliding wear behaviour of HVOF sprayed CoMoCrSi coating", *Materials Research Express* 5 (2018), 086519. https://doi.org/10.1088/2053-1591/aad44e.

[30] Y. C. Lin, S. W. Wang, "Wear behavior of ceramic powder cladding on an S50C steel surface", *Tribology International* 36 (1) (2003), 1–9.

Chapter 7

Processing Functionally Graded Materials through Microwave Heating

Sarbjeet Kaushal

Contents

7.1 Introduction to Functionally Graded Materials

In today's world, industries prefer materials that exhibit multiple properties under different working conditions. Wear problems are faced by many industries all across the world. The wear phenomenon causes financial losses to these industries. Surface engineering is the solution to counteract the wear problem faced by these industries. Surface engineering deals with the modification of the surface of material vulnerable to various problems such as corrosion, wear, and so on. Single-layer metal ceramic–based composite claddings/coatings are commonly used surface engineering techniques now these days. However, in metal ceramic–based composite clads, high transverse shear stresses and inplane normal stresses at the interface occur due to the presence of large differences in the properties of metal and ceramic-based materials. These stresses at the interface cause the delamination/decohesion at the interface of the metal and the ceramic, resulting in the poor load-bearing capacity. The problem of a sharp interface can be avoided using materials that replace this sharp interface with layers of gradual varying composition and microstructure. Such materials are known as functionally graded materials.

DOI: 10.1201/9781003248743-7

Functionally graded materials (FGMs) are widely used in bioengineering, power generation, aerospace, automotive, and structural demand applications. FGMs offer many advantages over conventional composite and alloy materials. FGMs give the advantage of taking the benefits of different materials' properties such as ceramics and metals. FGMs can be produced by different composition materials such as metal–ceramic, metal–metal, ceramic–polymer, and ceramic–ceramic. There are many fabrication techniques available for FGMs such as centrifugal casting, physical vapor deposition, solid freeform fabrication, selective laser sintering, laser-engineered net shaping based on powder deposition, plasma spray forming, and so on. However, in recent years, researchers are exploring material processing methods that are sustainable and provide the processing of materials at lower processing time with lower energy consumption. Research in the field of microwave processing of materials has emerged in recent years, owing to its unique advantages over conventional material processing techniques. Microwave material processing methods offer volumetric heating property, which is the noble characteristic of microwave heating. The volumetric heating characteristics of microwave heating provide the uniform thermal gradient in the processed materials and result in fewer stresses in the materials. Many researchers concluded that microwave heating can process the materials in less processing time and consume less power as compared to the conventional material processing methods. Srinath et al. [1], in 2011, produced the microwave joining and reported that the microwave processed joint exhibits excellent tensile strength. Gupta et al., in 2011 [2], [3], extended the work of microwave heating in the field of microwave cladding of metallic and metal-ceramic–based composite materials. The authors reported that microwave processed claddings exhibited excellent wear resistance properties. Further authors [4], [5] reported the successful development of metal-ceramic–based composite castings through microwave heating. Processing of metals and metal-ceramic–based materials encouraged the researchers to work on the development of functionally graded materials through microwave heating. In this chapter, the work performed by various research on metal-ceramic-based FGMs through microwave heating has been reported.

7.2 Processing of Various FGMs through Microwave Heating

7.2.1 Microwave Sintering of FGMs

High-temperature applications of microwave heating made the research community think about the processing of high-temperature materials through it. In 2002, a group of scientists at Pennsylvania State University successfully fabricated the commercial powdered-metal gear parts through microwave sintering technique within 30 min of exposure time [6]. In this study authors also reported that the magnetic field also plays a crucial role in the microwave sintering of bulk semiconductor and conductor materials. This research paved the path for the researchers to work in the domain of microwave sintering of metal-ceramic–based materials. Microwave heating is associated with several advantages such as rapid heating, eco-friendly nature, improved quality of product over conventional heating methods [7]. Authors [7] processed various materials such as ceramics, composites, metals, and glass through the microwave heating route. It was claimed in the study that microwave sintered ceramic exhibited better properties than conventional sintered ceramics. The alumina-based ceramic sample was sintered through microwave sintering at 200°C in less time than with the conventional sintering method and that, too, at a lower processing time. The density obtained during microwave sintering was almost 98%. Microwave-sintered samples showed 50% more hardness than conventionally sintered samples. Afterward, a lot of studies were carried out in the domain of microwave sintering of various kinds of materials.

The concept of microwave sintering has been further explored in the domain of the development of functionally graded materials. In 2002, Katakam et al. [8] fabricated the microwave processed bioactive FGM by gradually varying the composition of calcium phosphates from the surface to the interior. The FGMs were formed by both conventional and microwave sintering methods. The authors reported that the greater composition control was present in FGM fabricated through microwave processing. Liu et al., in 2012 [9], reported the successful sintering of microwave processed W/Cu-based FGM. It was reported in the research that the five-layered W/Cu based FGM with different compositions of W and Cu (W30% + 70% Cu, W50% + 50% Cu, W70% + 30% Cu, W90% + 10% Cu, and W100%) were developed through microwave sintering within 30 min of microwave exposure. The microstructural analysis through scanning electron microscopy (SEM) images and energy dispersive X-ray spectroscopy (EDS) maps revealed that the graded structure of Cu/W-based samples can be retained after microwave sintering (above 1300°C temperature). The thermal conductivity of the FGM sample was observed as 200 W/mK at ambient temperature. Figure 7.1 shows the X-ray diffraction (XRD) pattern of functionally graded clad (FGC) layers and a photographic view of FGC samples.

Tang et al. (2016) [10] reported the formation of cemented carbide-based FGMs through microwave-assisted nitriding sintering as shown in Figure 7.2. The authors studied the effect of sintering temperature and composition on the phase composition, microstructure, and mechanical properties of FGMs. Two functionally graded cemented carbide samples named FGCC-T5 (WC5TiC10Co) and FGCC-T15 (WC15TiC6Co) were prepared. The green compacts were made using powder metallurgy methods followed by heating in the vacuum sintering furnace at 400°C to remove the wax. Experiments were performed in a microwave oven at 2.45 GHz. The hardness and

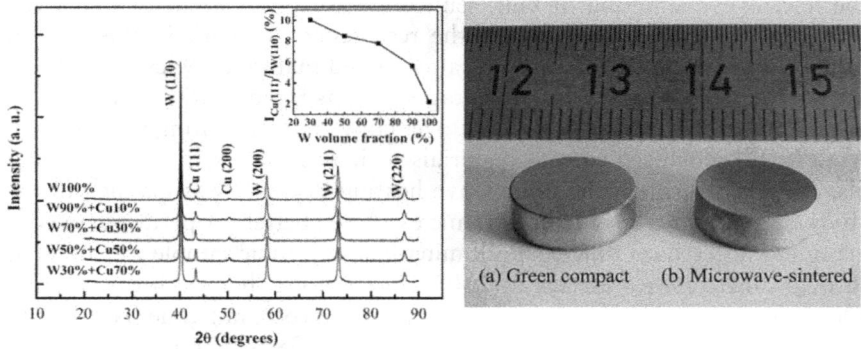

Figure 7.1 (a) XRD patterns for different microwave-sintered layers; (b) photographic view of a W/Cu-based green compact and microwave-sintered FGC sample [9].

Source: Reused with permission of the publisher.

strength of FGCC-T5 FGCC were observed as 1580 HV and 2010.2 MPa at 1400°C, while FGCC-T15 exhibited the hardness and strength of 1835 HV and 1560 MPa.

In 2012, Bykov et al. [11] carried out the numerical modeling of microwave-sintered FGMs of Ni-Al$_2$O$_3$, Al$_2$O$_3$-NiAl, and Mo-ZrO$_2$ based materials (Figure 7.3). Numerical calculations of densification of FGMs have been done based on sintering kinetic data of pure metals. The porosity percentage of FGM sample layers have been shown less than 5% except for the layers with presence of more than 80% metal content.

Microwave sintering of four-layered diamond/W-Cu-based functionally graded materials was reported by Wei et al. [12] in 2021. The authors concluded in their research that the density of diamond/W-Cu-based FGC at 1200°C sintering temperature was 96.6%, with thermal conductivity of 220 W/mK. An EDS analysis of FGM confirmed the presence of different functionally gradient layers in microwave-sintered FGM at various locations.

Microwave sintering of W/Cu-based FGM was reported by Zhou et al. [13] in 2018, as shown in Figure 7.4. The authors have designed a microwave sintering setup to provide a significant thermal gradient and potent heat flux during sintering. In this work, W skeletons were synthesized first using microwave sintering followed by infiltration with Cu to form Cu-W-based FGM. The gradient in porosity and hardness was reported by the authors. Furthermore, the authors revealed that the composition and microstructure of FGM can be tailored by adding Ni as a sintering activator.

In 2017, Bharathi et al. [14] investigated the mechanical properties of aluminum and silicon carbide–based FGMs processed through a microwave

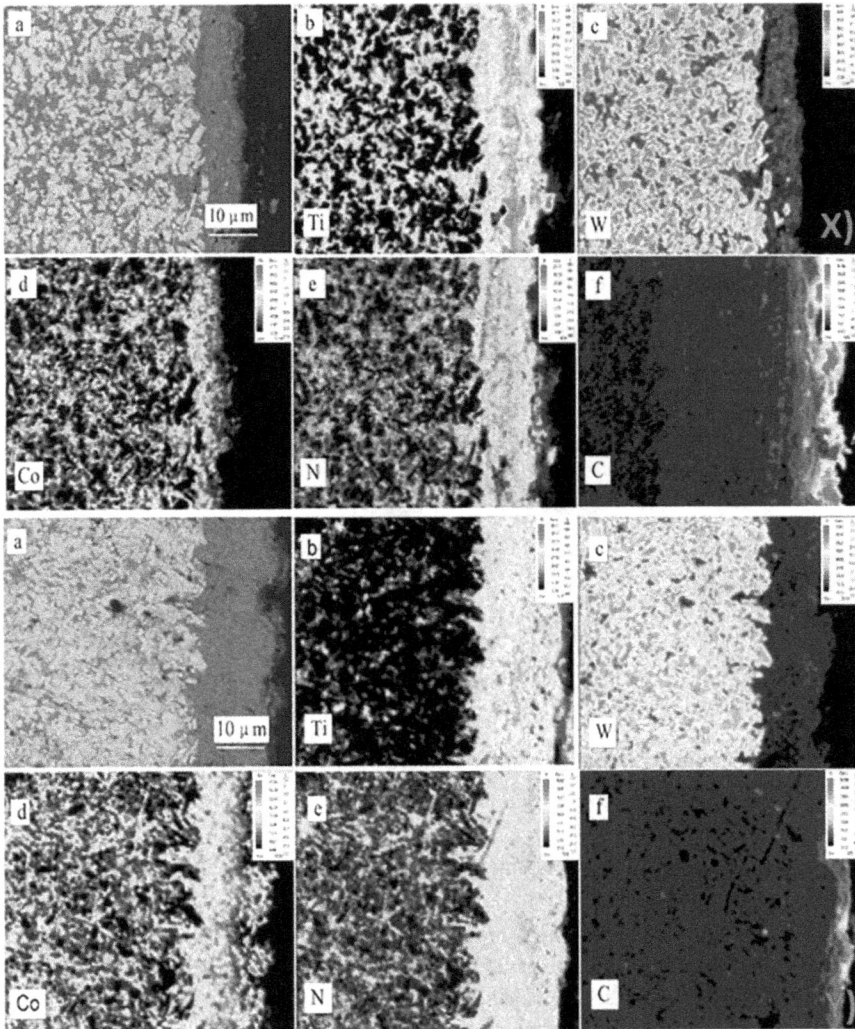

Figure 7.2 (x) Typical cross-sectional microstructure of FGCC-T5, (y) FGCC-T15 [10].

Source: Reused with the permission of the publisher.

sintering route as shown in Figure 7.5. It was reported that the increase in the percentage of SiC particles along the deposition direction resulted in higher porosity level and hardness of FGMs. Different values of hardness were observed at the different ends of FGMs due to differences in the SiC

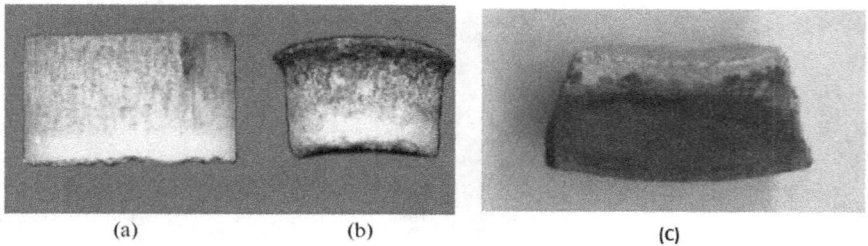

Figure 7.3 Photographs of transverse section of NiAl-Al$_2$O$_3$ multilayer graded samples (a) before; (b) after sintering; (c) transverse section of sintered multilayer graded Mo-ZrO$_2$ sample [11].

Source: Reused with permission of the publisher.

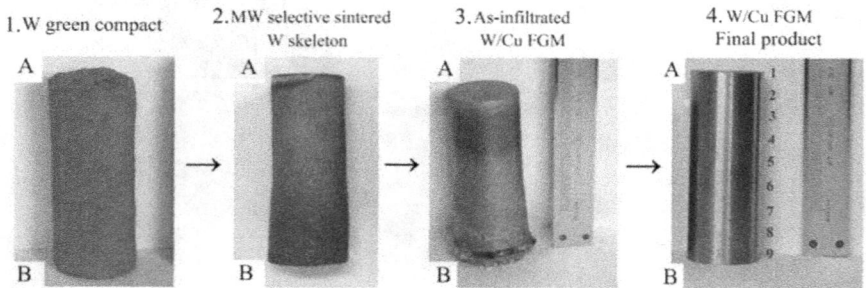

Figure 7.4 Different stages for formation of microwave-processed W/Cu-based FGM [13].

Source: Reused with permission of the publisher.

volume fraction at both ends. Thermal tests confirmed the effective bonding between the different layers.

Bykov et al. [15] developed a method for sintering if metal-ceramic FGM through millimeter-wave gyrotron heating as shown in Figure 7.5. Six layered FGMs of Ni-Al$_2$O$_3$-based materials have been developed by powder stacking and consolidation by millimeter-wave sintering. The authors reported that the developed method can be used for developing the FGMs for different compositions of dissimilar materials having different sintering temperatures.

7.3 Microwave-Processed FGCs

The emergence of microwave material processing in the microwave claddings [16]–[21] domain encouraged the researchers to work in the development of FGCs through microwave heating process.

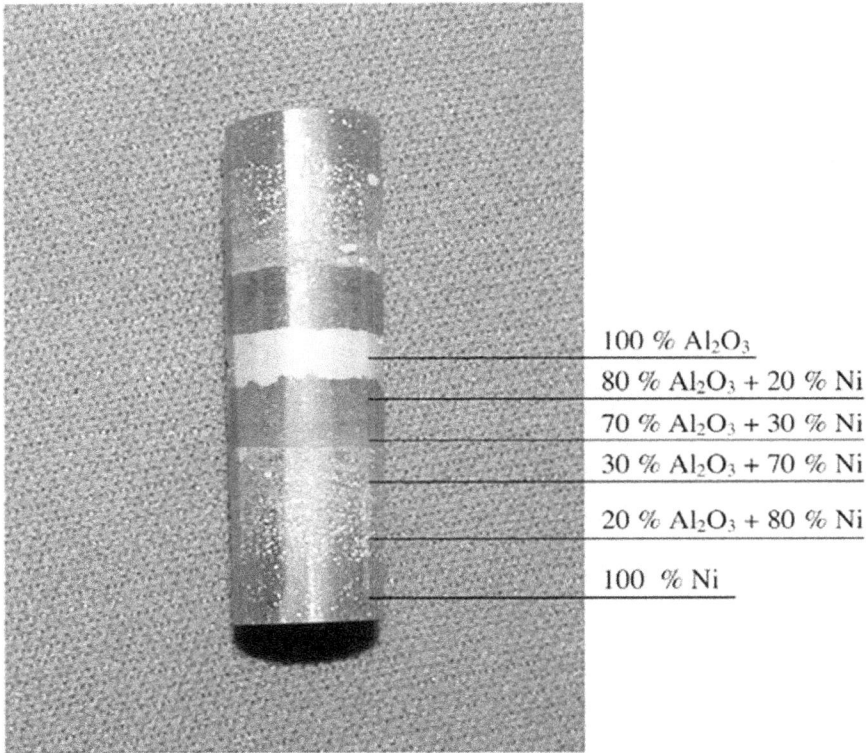

100 % Al₂O₃

80 % Al₂O₃ + 20 % Ni

70 % Al₂O₃ + 30 % Ni

30 % Al₂O₃ + 70 % Ni

20 % Al₂O₃ + 80 % Ni

100 % Ni

Figure 7.5 Photographic view of microwave sintered Ni-Al₂O₃-based FGM [15].

Source: Reused with permission of the publisher.

7.3.1 Mechanism of Formation of Functionally Graded Materials through Microwave Heating

The formation of functionally graded materials through microwave heating is a challenging job due to the reflection of microwaves by metal particles at room temperature. However, the interaction of metal powder particles with microwaves can be made by using the microwave hybrid heating (MHH) route. Authors [22] reported the principle of MHH in detail. The systematic of MHH is shown in Figure 7.6a. The heating phenomenon inside the microwave oven is shown in Figure 7.6b, which revealed the red-hot heating profile of the charcoal powder during microwave heating. Figure 7.6c revealed the steps used for producing microwave FGCs.

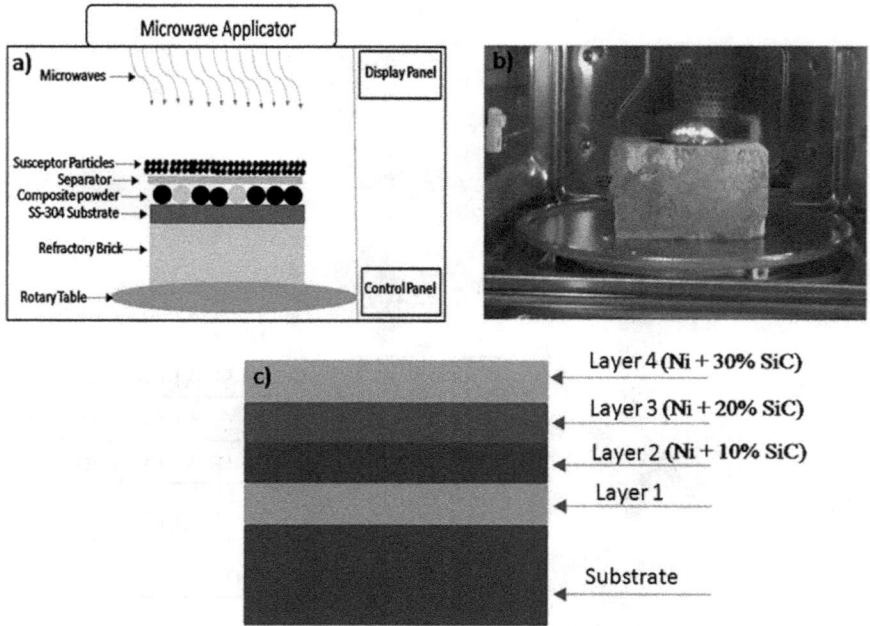

Figure 7.6 (a) Mechanism of formation of FGC layer through microwave irradiation; (b) heating inside microwave oven due to microwave irradiation; (c) systematic arrangement of different FGC layers in a stepwise manner [23].

Source: Reused with permission of the publisher.

7.3.2 Various Research Done in the Domain of Microwave Processed FGCs

In 2018, Kaushal et al. [24] successfully practiced the feasibility of the formation of functionally graded composite clads through microwave heating mode. Four layers of Ni-WC-based materials with varying composition of WC-based ceramic powder from 0–30% were developed on austenitic stainless steel substrate using stepwise method of development of FGC. The FGC of approximately 1.8 mm thickness was developed with no interfacial cracks and minimal porosity percentage. The formation of various hard phases such as NiSi, $Cr_{23}C_6$, NiW_4, and others was reported during the microwave heating, resulting in the higher value of microhardness of FGC layers.

Furthermore, authors [23], [25] have reported the formation of FGCs of Ni-SiC-based composite material on a stainless steel substrate through microwave energy as shown in Figure 7.7. Four layers FGC with varying SiC content from 0–30% with a step size of 10 wt.% were produced. An FGC with a thickness of 2 mm was achieved at 2.45 GHz and 900 W. It

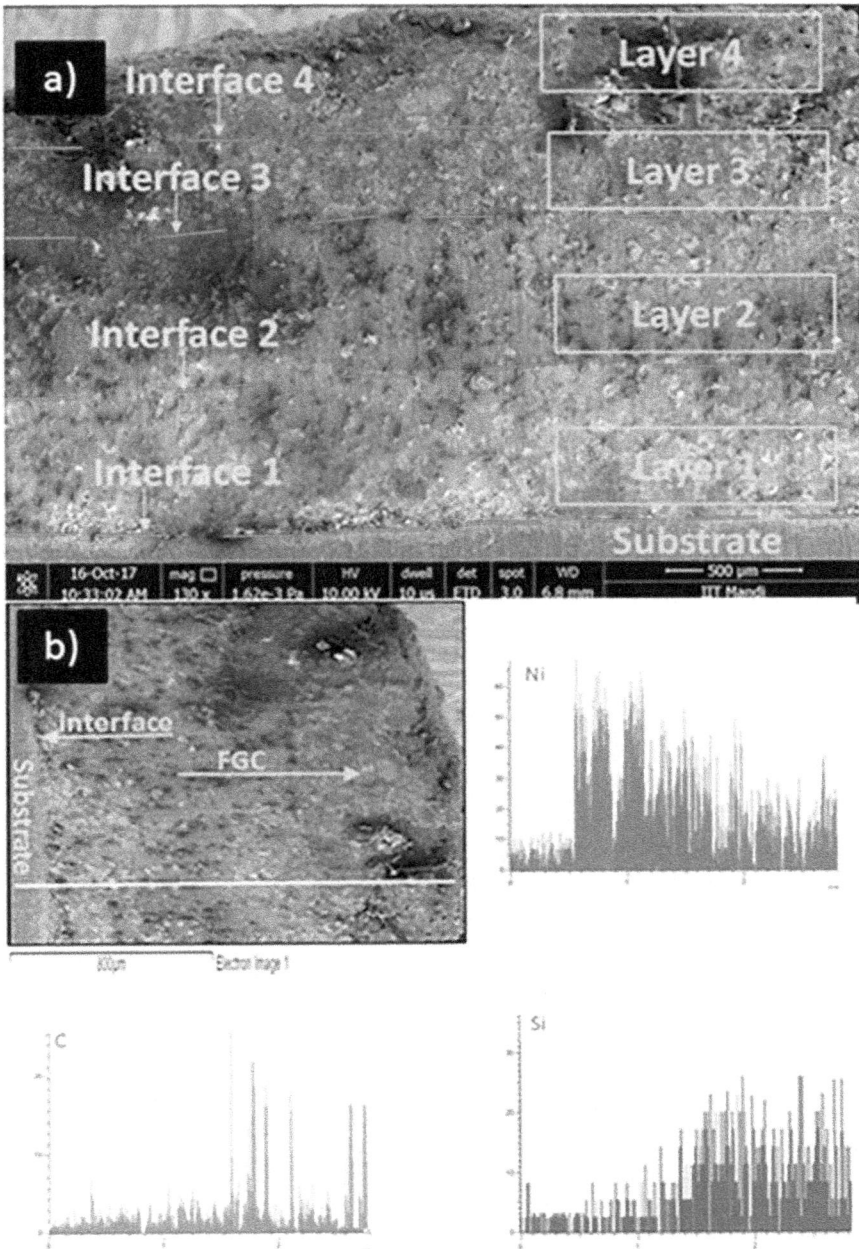

Figure 7.7 SEM and EDS line mapping for a microwave-processed Ni/SiC-based FGC [23].

Source: Reused with permission of the publisher.

was observed that the top FGC layer exhibited the maximum microhardness value of order 1020 ± 30 HV due to the presence of hard SiC particles. The presence of various phases such as $FeSi_2$, $NiSi$, Ni_3C, Si, $Cr_{23}C_6$, and C_3Cr_7 were observed in different FGC layers. The presence of hard phases in FGC layers made it useful for wear-resistance applications. The presence of varying ceramic phases in the Ni-based matrix of different layers resulted in varying hardnesses of the FGC layers. It was reported that the upper FGC layers of various FGC showed a higher value of microhardness. Table 7.1 shows the comparison of various microhardness values exhibited by upper FGC layers for different FGC materials. It was observed that Ni-SiC-based FGC upper layer exhibited a maximum value of microhardness among other FGC layers. It might be due to the higher inherent hardness of SiC particles compared to Cr_3C_2 and WC particles.

Authors [27] studied the tribological behavior of microwave processed functionally graded clads of $Ni-Cr_3C_2$ based materials. Four different layers of $Ni-Cr_3C_2$-based materials (with the varying proportion of Cr_3C_2-based powder from 0–30% in a step size of 10% by weight) were fabricated over SS-304 substrate through microwave irradiation at 2.45 GHz. Wear study of the microwave processed FGC was carried out using pin-on-disk tribometer against a counter disc of alumina. The wear study was performed under various process parameters, such as sliding distances, sliding velocity, and normal load. It was reported that FGC exhibited excellent wear resistance compared to single-layer clads and substrates. Table 7.2. illustrates the wear loss in case of various microwave processed FGC under 1.5 kg of normal load, 1 m/s of sliding velocity and after 2000 m of sliding. It was observed that Ni-SiC-based FGC performed best among all other FGMs. The higher microhardness of the Ni-SiC-based FGC was responsible for its better wear resistance.

Table 7.1 Comparison of Microhardness Values of Various Microwave-Processed FGCs

S. No	Material	Maximum Microhardness (HV)
1.	Ni-WC based FGC [26]	880 ± 30
2.	Ni-SiC based FGC [25]	1025 ± 30
3.	$Ni-Cr_3C_2$ based FGC [27]	576 ± 25

Table 7.2 Wear Comparison of Various FGC at 1.5 kg, 1 m/s, and 2000 m

S. No	Material	Cumulative Weight Loss (mg)
1.	Ni-WC-based FGC [28]	2.4
2.	Ni-SiC-based FGC [23]	1.1
3.	$Ni-Cr_3C_2$-based FGC [27]	3.8

7.4 Conclusion

In this chapter, the various studies done in the domain of FGMs of various composite materials through microwave sintering and microwave cladding routes have been reported. The results of mechanical, metallurgical and tribological behavior of microwave-sintered and microwave-processed FGC were discussed. It was concluded that the microwave-processed FGMs exhibit excellent tribological properties under various sliding conditions. Furthermore, microwave heating results in the formation of FGC layers with excellent microstructure. Higher microhardness and wear resistance of microwave-processed FGCs make these suitable for wear applications.

References

[1] M. S. Srinath, A. K. Sharma, and P. Kumar, "Investigation on microstructural and mechanical properties of microwave processed dissimilar joints," *J. Manuf. Process.*, vol. 13, no. 2, pp. 141–146, 2011, doi: 10.1016/j.jmapro.2011.03.001.

[2] D. Gupta and A. K. Sharma, "Investigation on sliding wear performance of WC10Co2Ni cladding developed through microwave irradiation," *Wear*, vol. 271, nos. 9–10, pp. 1642–1650, 2011, doi: 10.1016/j.wear.2010.12.037.

[3] D. Gupta and A. K. Sharma, "Microstructural characterization of cermet cladding developed through microwave irradiation," *J. Mater. Eng. Perform.*, vol. 21, no. 10, pp. 2165–2172, 2012, doi: 10.1007/s11665-012-0142-2.

[4] S. Singh, D. Gupta, and V. Jain, "Novel microwave composite casting process: Theory, feasibility and characterization," *Mater. Des.*, vol. 111, pp. 51–59, 2016, doi: 10.1016/j.matdes.2016.08.071.

[5] S. Singh, D. Gupta, and S. Kaushal, "Dry sliding wear performance of Ni—SiC composites developed through an in situ microwave casting process," *J. Tribol.*, vol. 142, no. October, pp. 1–10, 2020, doi: 10.1115/1.4047032.

[6] J. Cheng, R. Roy, and D. Agrawal, "Radically different effects on materials by separated microwave electric and magnetic fields," *Mater. Res. Innov.*, vol. 5, nos. 3–4, pp. 170–177, 2002, doi: 10.1007/s10019-002-8642-6.

[7] D. Agrawal, "Microwave sintering of ceramics, composites and metallic materials, and melting of glasses," *Trans. Indian Ceram. Soc.*, vol. 65, no. 3, pp. 129–144, 2006, doi: 10.1080/0371750X.2006.11012292.

[8] S. Katakam, D. Siva Rama Krishna, and T. S. Sampath Kumar, "Microwave processing of functionally graded bioactive materials," *Mater. Lett.*, vol. 57, no. 18, pp. 2716–2721, 2003, doi: 10.1016/S0167-577X(02)01364-2.

[9] R. Liu et al., "Microwave sintering of W/Cu functionally graded materials," *J. Nucl. Mater.*, vol. 431, no. 1–3, pp. 196–201, 2012, doi: 10.1016/j.jnucmat.2011.11.013.

[10] S. Tang et al., "Microstructure and mechanical properties of functionally gradient cemented carbides fabricated by microwave heating nitriding sintering," *Int. J. Refract. Met. Hard Mater.*, vol. 58, pp. 137–142, 2016, doi: 10.1016/j.ijrmhm.2016.04.013.

[11] Y. V. Bykov et al., "Fabrication of metal ceramic functionally graded materials by microwave sintering 1," *Inorg. Mater. App. Res.*, vol. 3, no. 3, pp. 261–269, 2012, doi: 10.1134/S2075113312030057.

[12] C. Wei et al., "Fabrication of diamond/W—Cu functionally graded material by microwave sintering," *Nucl. Eng. Technol.*, no. xxxx, 2021, doi: 10.1016/j.net.2021.08.035.

[13] C. Zhou, L. Li, J. Wang, J. Yi, and Y. Peng, "A novel approach for fabrication of functionally graded W/Cu composites via microwave processing," *J. Alloys Compd.*, vol. 743, pp. 383–387, 2018, doi: 10.1016/j.jallcom.2018.01.372.

[14] R. Jayendra Bharathi and C. Suresh Kumar, "Microwave processing and mechanical property evaluation of functionally graded materials with Al/SiC powders," *Mater. Today Proc.*, vol. 5, no. 6, pp. 14481–14488, 2018, doi: 10.1016/j.matpr.2018.03.035.

[15] Y. V. Bykov et al., "Temperature profile optimization for microwave sintering of bulk Ni-Al2O3functionally graded materials," *J. Mater. Process. Technol.*, vol. 214, no. 2, pp. 210–216, 2014, doi: 10.1016/j.jmatprotec.2013.09.001.

[16] S. Kaushal, D. Singh, D. Gupta, H. Bhowmick, and V. Jain, "Processing of Ni20W-C10Mo based composite clads on austenitic stainless steel through microwave hybrid heating," *Mater. Res. Express*, vol. 5, no. 3, 2018, doi: 10.1088/2053-1591/aab0f2.

[17] S. Kaushal, D. Gupta, and H. Bhowmick, "Investigation of dry sliding wear behavior of Ni—SiC microwave cladding," *J. Tribol.*, vol. 139, no. 4, p. 041603, 2017, doi: 10.1115/1.4035147.

[18] S. Kaushal, D. Gupta, and H. Bhowmick, "On microstructure and wear behavior of microwave processed composite clad," *J. Tribol.*, vol. 139, no. 6, p. 061602, 2017, doi: 10.1115/1.4035844.

[19] D. Gupta, P. M. Bhovi, A. K. Sharma, and S. Dutta, "Development and characterization of microwave composite cladding," *J. Manuf. Process.*, vol. 14, no. 3, pp. 243–249, 2012, doi: 10.1016/j.jmapro.2012.05.007.

[20] B. Singh, S. Kaushal, D. Gupta, and H. Bhowmick, "On development and dry sliding wear behavior of microwave processed Ni/Al$_2$O$_3$ composite clad," *J. Tribol.*, vol. 140, no. 6, p. 061603, 2018, doi: 10.1115/1.4039996.

[21] S. Kaushal, B. Singh, D. Gupta, H. Bhowmick, and V. Jain, "An approach for developing nickel—alumina powder-based metal matrix composite cladding on SS-304 substrate through microwave heating," *J. Compos. Mater.*, vol. 52, no. 16, pp. 2131–2138, 2018, doi: 10.1177/0021998317740732.

[22] D. Gupta and A. K. Sharma, "Development and microstructural characterization of microwave cladding on austenitic stainless steel," *Surf. Coatings Technol.*, vol. 205, no. 21–22, pp. 5147–5155, 2011, doi: 10.1016/j.surfcoat.2011.05.018.

[23] S. Kaushal, D. Gupta, and H. Bhowmick, "On development and wear behavior of microwave-processed functionally graded Ni-SiC clads on SS-304 substrate," *J. Mater. Eng. Perform.*, vol. 27, no. 2, pp. 777–786, 2018, doi: 10.1007/s11665-017-3110-z.

[24] S. Kaushal, D. Gupta, and H. Bhowmick, "On processing of Ni-WC based functionally graded composite clads through microwave heating," *Mater. Manuf. Process.*, vol. 33, no. 8, pp. 822–828, 2018, doi: 10.1080/10426914.2017.1401724.

[25] S. Kaushal, D. Gupta, and H. Bhowmick, "An approach for functionally graded cladding of composite material on austenitic stainless steel substrate through

microwave heating," *J. Compos. Mater.*, vol. 52, no. 3, pp. 301–312, 2018, doi: 10.1177/0021998317705977.

[26] S. Kaushal, D. Gupta, and H. Bhowmick, "On processing of Ni-WC based functionally graded composite clads through microwave heating," *Mater. Manuf. Process.*, vol. 33, no. 8, pp. 822–828, 2018, doi: 10.1080/10426914.2017.1401724.

[27] S. Kaushal, D. Gupta, and H. Bhowmick, "Novel investigation on tribological behavior of microwave synthesized functionally graded claddings," *Proc., Inst. Mech. Eng., Part C: J. Mech. Eng. Sci.*, vol. 0, no. 0, pp. 1–10, 2021, doi: 10.1177/09544062211045892.

[28] S. Kaushal, D. Gupta, and H. Bhowmick, "Wear behavior of microwave-processed Ni-WC8Co-based functionally graded materials," *Proc. Inst. Mech. Eng. Part L J. Mater. Des. Appl.*, vol. 235, no. 5, pp. 1036–1045, 2021, doi: 10.1177/1464420720988119.

Chapter 8

Evolution and Adoption of Microwave Claddings in Modern Engineering Applications

Dinesh Kumar, Rahul Yadav, and Jashanpreet Singh

Contents

8.1 Introduction

At present, various industries throughout world face the need for the improvement in the material wear and corrosion-resistant properties that causes serious degradation of engineering devices, apparatuses, and machines. Surface protection of engineering materials is as significant as their strength when it comes to designing various components. The typical techniques of improving the surface characteristics of steel and its alloys include nitriding, carbonitriding, carburizing, coating/cladding, and others. The few techniques are discussed as follows:

- *Nitriding:* to improve the wear and fatigue resistance of steels
- *Carbonitriding:* to improve the wear resistance
- *Carburizing:* to improve the wear resistance and rolling-contact/bending fatigue
- *Physical vapour deposition (PVD):* to improve the wear (like tools and dies) and corrosion resistance and used for epitaxial growth of semiconductors

DOI: 10.1201/9781003248743-8

- *Thermal Spraying:* to improve the wear and corrosion resistance
- *Laser cladding process:* to improve the wear and corrosion resistance

Coating/cladding of the hard facing material is the key answer to such type of degradation problem. Cladding is the process in which a surface of a component is covered or coated with another material. Surface changes improve the hardness of the material and raise the material's resistance to wear and corrosion or erosion. The general techniques for the development of wear-resistant coats are thermal spraying, cold spraying, electroplating, sputtering, and laser claddings [1–3]. The produced coating has weak adherence with the base materials and poor resistance against pinpoint loading during thermal spraying. While the cladding produced by utilizing laser apparatus has much better strength and adherence due to metallurgical connection. However, the intensive point-heat source produces a heat distortion and significant residual stress in a laser clad. Moreover, the higher cooling rate in laser processing leads to flaws such as porosity and fractures in solidified coverings. Modern methods of processing like post-laser treatment, reheating of surfaces, and others can overcome laser cladding limitations and produce better mechanical and microstructures properties in a cheap way and at a higher processing speed. In the last few years, considerable work has been carried out in the area of microwave heating in order to develop wear-resistant reinforcement on metallic surfaces, which is discussed in the following.

8.1.1 Microwave Irradiation and Its Applications

The microwaves are part of electromagnetic spectrum having a wavelength in the range of 1 m to 1 mm and frequency lies in the range of 300 MHz to 300 MHz. The major advantage of employing microwaves is its reduced time and lower power consumption, which facilitates boosting the number of items produced. Microwave heating is entirely different from the conventional heating of materials. In microwave heating, the generation of heat depends on the dielectric property of the material being processed whereas conventional heating depends on the thermal conductivity of the material. Heating with microwaves happens at the atomic level by offering resistance by dipoles to the oscillating electric and magnetic fields components of the microwave energy. The atomic/molecular-level heating leads to homogeneous heating of the respective material, which improves the various mechanical properties of that material. Enhanced mechanical properties of the material can also be obtained due to fine the microstructures obtained during the process. As opposed to traditional approaches, this method also has less influence on the environment. There is a lot of study being done in this area since microwave processing does not have a negative impact on the environment. Microwave energy has various real-time applications and few applications, which are shown in Figure 8.1. Microwave energy can also be

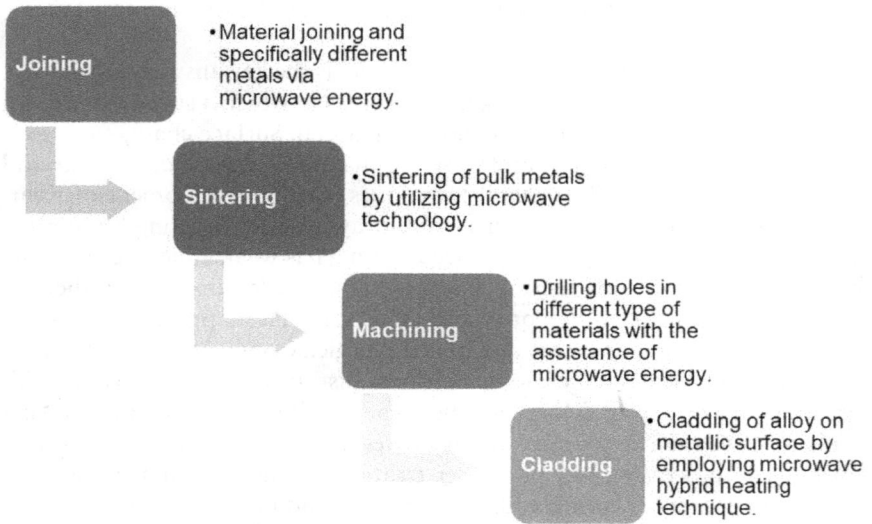

Figure 8.1 Microwave energy has various real-time applications and few applications.

utilized for treatment or processing of nonmetallic materials. Various examples for treatment of nonmetallic materials are found in applications like wood curing, polymer-based nanocomposites, medical waste processing, rubber tires devulcanization, nuclear applications, and food processing [4].

8.1.2 Historical Developments in Microwave Processing

Microwave energy was at first utilized in telecommunication sector. After the commencement of thrust research in the field of microwave energy, the scope of microwave energy was expanded to other applications such as food processing, waste management, processing of various materials (metallic as well as ceramic), steel making, rubber vulcanization, and the like [5–7]. The evolution of microwave energy in various applications has been illustrated in Figure 8.2. Before 1965, the role of microwave material processing was only in low-temperature applications (below 400°C). It wasn't until 1965 that microwave radiation was widely used in high-temperature applications. The first successful use of microwave energy in the field of material processing was in the sintering of materials using microwave radiation. Around 1999, the first literature on sintering metallic materials using microwave energy was reported, in which it was successfully shown that it is possible to couple metallic materials in powder form can be coupled with microwaves [8]. The sintering process took place by utilizing the absorption of

A
p
pl
ic
at
io

Te
le
co
m
m
un
ic

	Vulcani zation, Food process ing, pharma ceutical drying, wood/r ubber curing, paper drying	Bio che mis try, Pat hol ogy , Me dic al Sci enc e	Cer am ics and pro ces sin g tec hn olo gy	Non-oxide cerami cs (WC/ Co, SiC, Nitrid es, glasse s, etc. and proces sing techno logy	Met allic mat erial s and proc essi ng tech nolo gies	Coatin g/Met al joinin g techni ques and Claddi ng proces ses

<1950   1950   1960 | 1970   1980   1990   2000   >2000

Decades

< 400°C | > 500°C

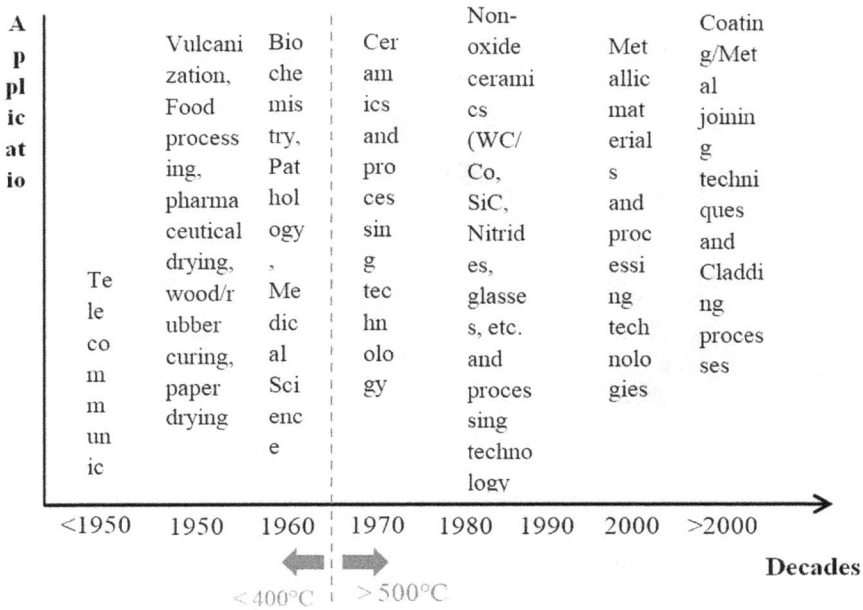

Figure 8.2 Developments in the field of microwave materials processing.

microwaves by ceramic powder. This development motivated researchers to focus on research and development in the field of microwave processing of various metallic powders. After the sintering process, microwave energy was continuously utilized in various operations like brazing operation, melting operation, coating operation, and joining operation [4].

Microwave cladding is the process of getting desired properties at the surface of one material by applying a layer of another material having superior properties with the help of microwave energy. Research in the field of microwave coatings/claddings and joining of metals picked up pace after 2009, when the need for better heat energy sources for surface coating of substrate metal and microwave energy was seen as an alternative for conventional heating.

8.2 Mechanism of Microwave Claddings

A typical schematic diagram of the fabrication of monolayer WC-10Co2Ni cladding by utilizing the concept of the microwave hybrid heating technique is shown in Figure 8.3. Gupta and Sharma [9] reported a detailed study on the mechanism involved in the microwave cladding process. According to

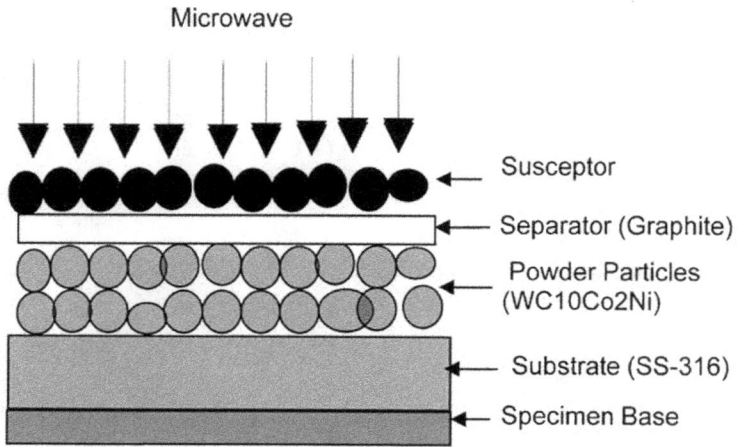

Figure 8.3 Typical schematic diagram of monolayer WC-10Co2Ni cladding [9].

their study, the *n* numbers of preplaced powder layers were poured on the surface at room temperature (T_r), as shown in Figure 8.4a. All the layers were at same the temperature before the microwave exposure. However, the interfacial temperature was assumed same as room temperature and denoted as T_{ina}. After applying the microwave irradiation, the susceptor is heated and transmits heat to the separator by the combined effect of conduction and radiation heat transfer. Conventional heat flow through the separator first raises the temperature of the powder particles' upper layers. A rise in skin depth at high temperatures causes EWAC particles to absorb microwave radiation, as illustrated in Figure 8.4b. All the preplaced powder layers begin to interact or couple with microwave radiation after a certain amount of heat has transferred from the upper layers to lower levels. To put it another way, powder particles begin to melt, causing the substrate-layer temperature to rise close to its melting point, and the individual powder layers begin to merge with each other, as shown in Figure 8.4c. A scanning electron microscopy (SEM) image of a typical EWAC-microwave cladded cross section is shown in Figure 8.5. From literature, it can be observed that the cladding produces the pore-free surface as compared to the spray coatings. The low porosity in the microwave induced clad products was due to uniform and volumetric heating phenomenon associated with microwave heating process.

8.2.1 Applications of Microwave Claddings

Numerous engineering applications require materials with high resistant to chemical oxidations and wear and which have ability to function

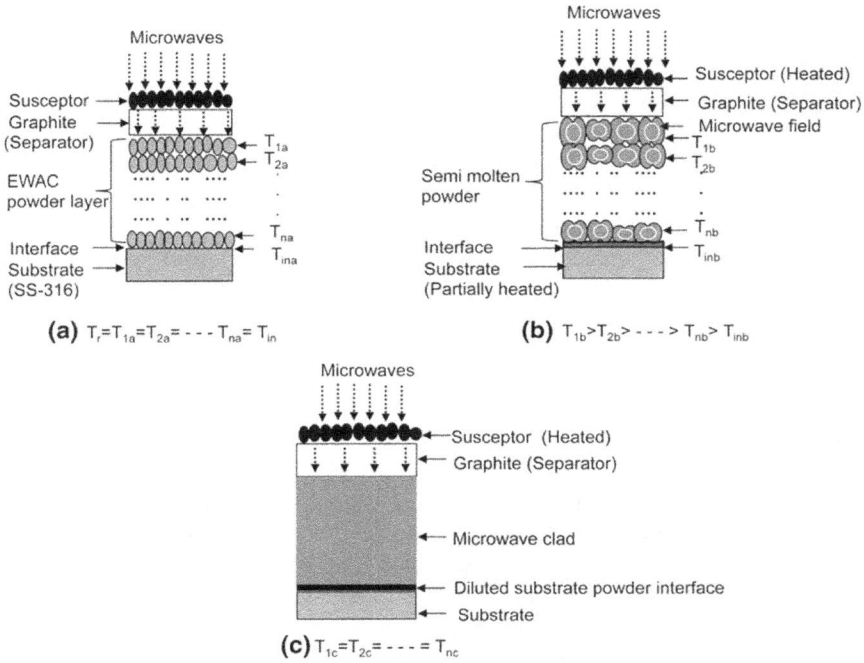

Figure 8.4 Schematic diagram of the microwave cladding process: (a) initial condition, (b) melting during microwave absorption, and (c) formation of clad layers [9].

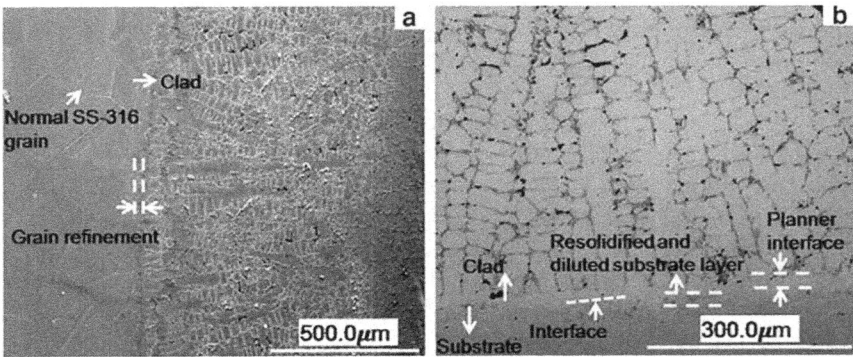

Figure 8.5 (a) SEM image of EWAC-cladded cross section and (b) back-scattered SEM image indicating the cellular structure [9].

properly even in harsh and hazardous conditions. Marine, aviation, oil and gas refineries, drilling, agriculture, and defense are various fields, which include both severe conditions and harsh environments. Most of the applications mentioned earlier require materials to stay in conditions susceptible to corrosion and other surface defects, which affects the performance of materials during operation. The materials employed in these applications also need to be strong, hard, and robust enough to handle the different structural and thermal stresses they have. To face these challenges materials surface properties are improved with the help of various conventional techniques (nitriding, carburizing, cyaniding, etc.) and microwave cladding or surface coating. The thermal scraping, high-speed oxy-fuel spray, tungsten inert gas (TIG), laser covering, and submerged metal arc-welding (SMAW), among others, are suitable ways of surface modification. Laser cladding is the most effective way to decrease the cracks, porosity, and bonding defects along with the controlled cooling rate. However, the main problem with laser procedures is the expense and thermal strains created by the heat levels [10]. There are different types of materials used to protect the different engineering applications, which are listed in Table 8.1.

Recently, the microwave processing of materials is gaining popularity due to its numerous advantages such as uniform and volumetric heating, environmentally friendly, and fewer residual stresses due to the low thermal gradient associated with this process. Furthermore, many materials are used for microwave cladding, such as metals, alloys, ceramics, and so on. WC-Co-Ni might be one of the finest solutions for high toughness and durability combinations. Tungsten carbide (WC) has higher hardness and wear strength but lower toughness. However, WC provides improved wear resistance by bonding in the tough phase. Cobalt (Co) functions as a binder in this system, which is responsible for the densification of fluid sintering by wetting, spreading, and agglomeration [11, 12]. WC10Co2Ni clad powder is utilized in various tribological components utilized in the gas turbine power plant as well as hydro-power plant [13]. The microwave-produced WC-12Co clads may be employed in various wear-resistant applications (automobile components, power generation units, aircraft engines, etc.) efficiently. Furthermore, the both micrometric and nanometric WC-12Co clads over austenitic stainless steel may also be used in applications like rock drilling, ore crushers, dies in powder metallurgy, and the like [14, 15]. Nickel-based clad powder is used in various engineering devices, such as hydraulic turbines, gas turbines, pumps, boilers, and so on. Various composite nickel-based powders (EWAC + 20% WC10Co2Ni, Ni-WC) are utilized in various engineering industries where a wear-resistant property is required during the operation. EWAC + 20%WC10Co2Ni can be used over mild steel in various applications due to poor friction and wear properties of

Table 8.1 Various Applications of Various Types of Materials Used to Protect the Different Engineering Components Utilized in Various Applications [4]

Substrate Material	Coating Powder	Application
Mild steel	EWAC + 20% WC10C2Ni	Diverse engineering firms produce various components as part of their scope of work
Austenitic Stainless Steel (SS304)	Micrometric and nanometric WC-12Co	Industrial applications such as rock drilling, ore crushers, earth-moving equipment, dies in powder metallurgy, and extruders, etc. Structural and machine components
Austenitic steel (SS-316L)	Ni (matrix)-80 wt.%WC (reinforcement)-20 wt%	Wear-resistance applications (power plants, pumps, ash conveying systems, turbines, etc.)
Martensitic stainless steel (AISI-420)	Alloy of cobalt (EWAC-1006EE)	Hydraulic and gas turbine components
Stainless steel (SS-304)	WC-12Co	Agricultural machinery, rock crushers, earth-moving and mining equipment, etc.
Austenitic stainless steel (SS304)	Micrometric and Nanometric WC—12Co	Engineering components such as in petrochemical, power and aerospace
Austenitic stainless steel (SS316)	WC10Co2Ni	Tribological components in gas turbine plant and hydro-power plant
Stainless steel (AISI 304)	WC-12Co	Engineering components in automobiles, power generation units, aircraft engines, etc.
High-speed steel (austenitic steel; SS-304)	Alloy of nickel (Cr-0.17, C-0.2, Si-2.8, Ni-Bal)	Components required different properties at different locations of the component
Austenitic stainless steel (SS-316)	EWAC (Ni-based) + 20% $Cr_{23}C_6$	Applications which require higher wear resistance
Austenitic stainless steel (SS-304)	Alloy of nickel	Hydraulic turbines, gas turbines, pumps, boilers, and fluid transport systems
Stainless steel (316L)	Mixtures of Ni and SiC	Fluid machines operating under extreme work conditions

mild steel [16, 17]. There are several fluid machines that employ Ni-SiC combination, which is used in severe working circumstances. Its high hardness and superior wettability with Ni lead to the selection of SiC as a material [18]. Co-based alloys are often utilized as a layer deposition material because of their good strength, corrosion, and erosion resistance at very high temperatures. Cobalt (EWAC-1006EE)-based alloys are used in hydraulic and gas turbine components [19].

8.3 Erosion-Resistant Claddings

Several factors influence the erosive process, including particle character-istics, target material qualities, and technique parameters, as mentioned in Table 8.2. Finne [20] said that the initial hardness of a material determines its failure mode. Brittle materials are often damaged by the cutting process, whereas ductile materials are typically damaged by plastic deformation. At the beginning of the erosion studies, Levy [21] found that ductile materials were more prone to erosion that decreased with time. Bellman and Levy [22] have outlined the mechanisms by which macroparticles generate erosion. Macroscopic particles are responsible for the formation of shallow craters and platelets. Deformation, ploughing, cracking, lip formation, and tiny cra-ters all occur in ductile materials [23]. Among other things, they asserted that the fast impingement of particles causes the material's surface to break down first. Figure 8.6 shows the schematic diagram for the erosion of two different types of materials.

Table 8.2 Slurry Variable Parameters Influence the Erosion Wear Rate [24].

S. No.	Slurry Variables		Material Variables	
	Type of Property	Variables	Type of Property	Variables
1	Liquid	Density, viscosity, lubricity, surface activity, pH, and temperature	Bulk properties	Brittleness/ductility, hardness/toughness, melting point, microstructure, shape/size, and roughness
2	Particles	Ductility, size, density, relative velocity (speed), impact angle, shape, relative hardness or hardness ratio, concentration, and particle/particle interactions	Surface properties	Surface treatment, work hardness, corrosion film, coating technique or method, powder composition, and grain size (microstructure)
3	Flow field	Angle of impingement, collision efficiency, viscosity, boundary layer, particle rebound, particle dropout, and turbulence intensity	Service variables	Contacting material, pressure, kinetic energy temperature, surface-roughness, lubrication, corrosion, hydraulic geometry, and flow

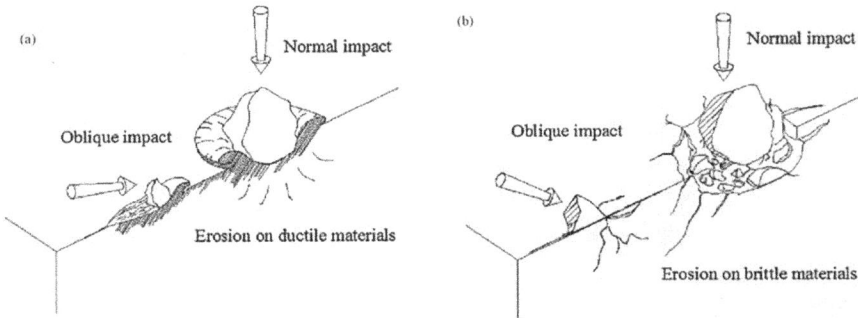

Figure 8.6 Erosion mechanism in the (a) ductile and (b) fragile materials [23].

Singh and Zafar [25] developed Ni+xCr$_7$C$_3$ composite cladding and tested their slurry erosion performance. Ni+30%Cr$_7$C$_3$ microwave composite cladding was found better over Ni+20%Cr$_7$C$_3$ and Ni+10%Cr$_7$C$_3$. Babu et al. [26] developed an erosion-resistant Ni-5%SiC microwave composite bimodal cladding. They reported that there were no pores in the claddings' metallurgical interface with the substrate. They achieved at least two times greater erosion resistance than hydro-turbine steel at all impact angles by cladding with bimodal Ni-5%SiC claddings. Hebbale and Srinath [19] tested the erosion mechanisms on the Co-based EWAC-1006EE claddings at different impact angles, which are shown in Figure 8.7. A comparison of erosion rates in terms of mean depth by depositing different types of laser cladding (LC) and microwave-cladded (MC) Ni-SiC composites is shown in Figure 8.8.

8.4 Corrosion-Resistant Claddings

Corrosion is the chemical deterioration of materials such as metals, insulators, semiconductors, polymers, and so on caused due to exposure of these materials to the surrounding environment. The deterioration of metals or metallic alloys occurs due to electrochemical reaction or chemical oxidization in surrounding environment [28]. Corrosion is slow, gradual, and natural phenomenon completely dependent on the atmospheric conditions, to which functional material is exposed [29]. Corrosion is caused by an electrochemical reaction that causes the material to degrade away from the target surface. These would be the two main processes of corrosion namely pitting and crater, respectively [30]. Corrosion is one of the factors that limit life span of various devices being used in engineering applications. Corrosion not only causes the decrease in efficiency of various devices, such as heat exchangers, but also causes deterioration of other properties like

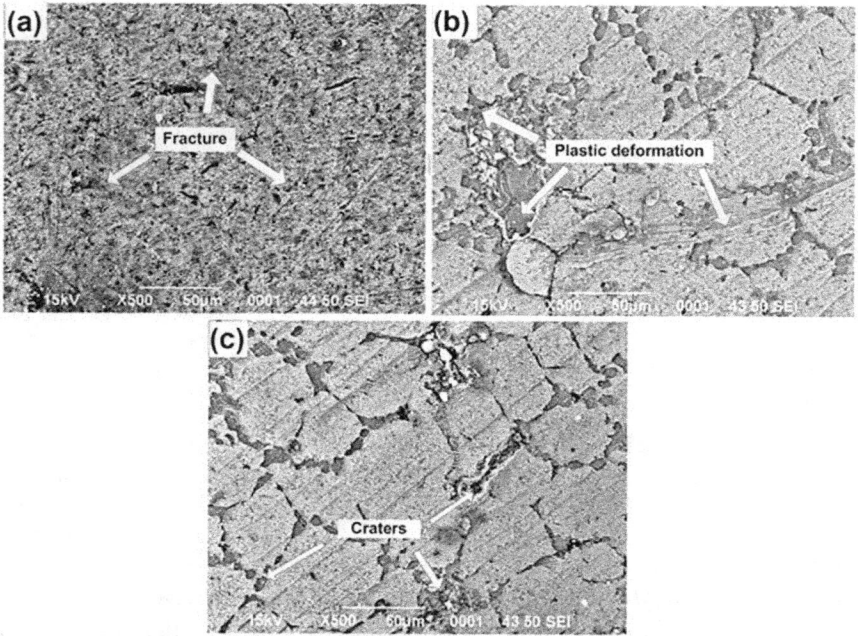

Figure 8.7 Erosion mechanisms in the Co-based EWAC-1006EE cladding at (a) 15°, (b) 30°, and (c) 45° [19].

appearance, strength, roughness, and so on of functional material. Since corrosion is a surface phenomenon, the effect of corrosion can be reduced by coating the functional metal surface with appropriate coating or cladding, that is, metallic (tin, zinc, etc.) or nonmetallic (porcelain enamels, glass coating, etc.), inhibitors (chromates, silicates etc.), and so on. Extensive utilization of cladding over functional metal surface to protect it from deterioration is a very old technique.

But the use of microwave energy in the cladding process, that is, microwave cladding, is a relatively new and better approach for developing cladding over functional substrate material [31]. The use of microwave energy has the advantages of low power consumption, less processing time and rapid heating, uniformity, and volumetric heating of substrate material [31].

Aluminum and Al-SiC are widely utilized cladding material in various industrial applications. Al has properties like high corrosion resistance, light weight, high strength, high reflectivity, and nontoxic while SiC is a high-microwave-absorbing material, making them suitable for cladding applications. These cladding materials can be used over substrate material such as mild steel. Since corrosion is completely affected by the surrounding

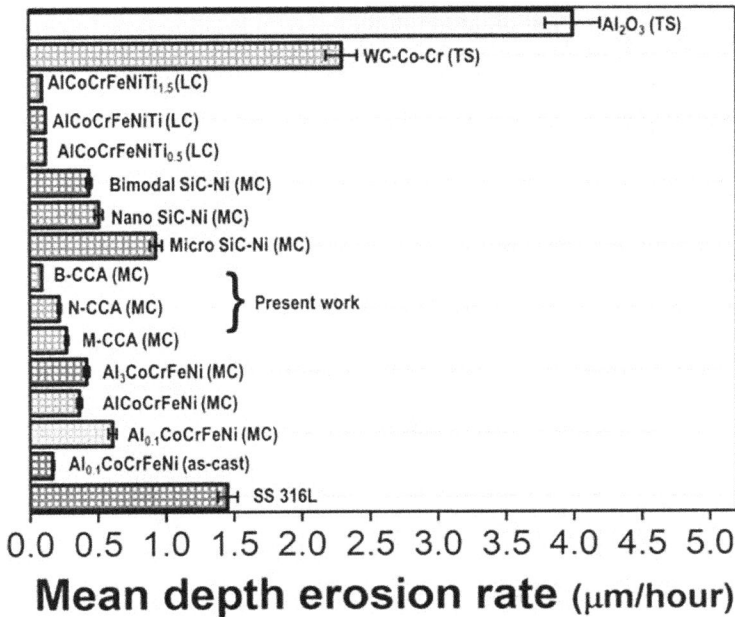

Figure 8.8 Erosion rates of LC, thermal sprayed (TS) coatings, as-cast CCA, and microwave-cladded (MC) Ni-SiC composites [27].

environment, it was found in study that corrosion proceeds at much faster rate in aqueous solutions than in acidic solutions [31]. The addition of hard ceramic and creation of an oxide layer on substrate surface are both responsible for alleviating lower corrosion rates in an acidic environment. In the case of an aqueous or a chloride environment, the cladding material obstructs the movement of ions, leading to minimal material loss or decay.

For corrosion prevention, anodic protection or the development of a physical barrier between metal and the environment may be achieved by using polymer-cladded metals such as steel [33, 34]. Anodic protection involves cladding of base metal with conducting polymeric compounds, such as polypyrrole, polyaniline, polythiophene, and others. Microwave-assisted polyester cladding is used over steel metal to improve its corrosion resistance by creating physical barrier between steel and surrounding. This type of cladding provides uniform coating thickness over the substrate metal, that is, steel. The process of utilizing microwave energy for polyester polymer cladding on cold rolled annealed steel is shown in Figure 8.9. Loharkar and Ingle [32] tested the mentioned polyester cladding for corrosion resistance by subjecting coating to a natural salt-spray test at 35 + 1°C with an NaCl solution

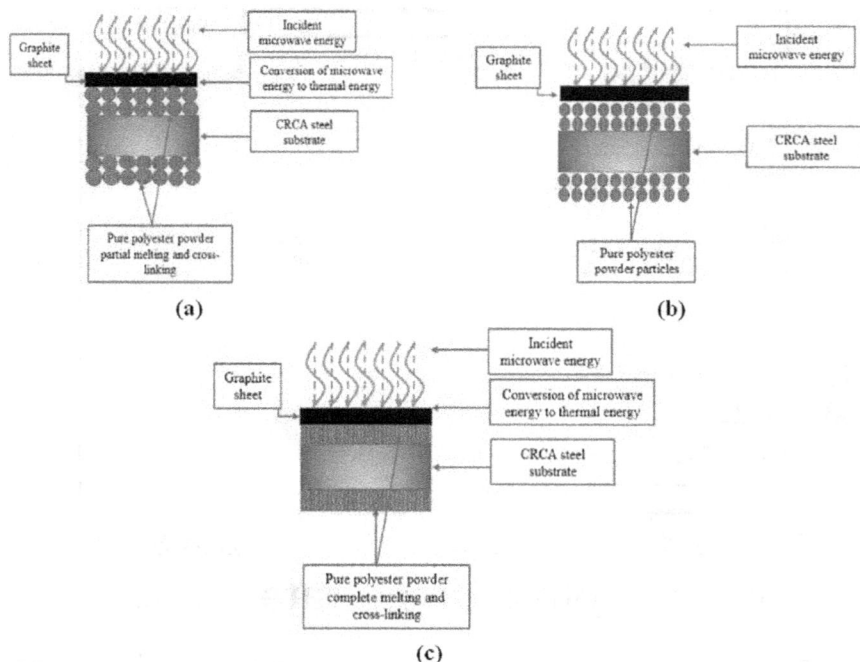

Figure 8.9 Schematic process of polyester powder cladding using microwave energy: (a) initial stage: microwaves incident on graphite sheet, (b) partial melting and initiation of cross-linking, and (c) melting and curing of powder [32].

in demineralized water for a duration of 152 hours. No adhesion failure or blister was absorbed during the testing period, which proves the effectiveness of polyester cladding in reducing the effects of corrosion.

8.5 Thermal Barrier Claddings

The thermal barrier claddings are generally applied to the high-temperature oxidation or hot corrosion environment. High-temperature oxidation is a corrosion process that takes place in metals when they come in contact with hot gases at elevated temperatures. The purpose of these claddings is to provide resistance to the oxidation in extreme temperature conditions, such as boilers, aircrafts' or airplanes' outer surface and its engine components, nuclear power plants, thermal power plants, and others. High-temperature corrosion, or hot corrosion, is one of the major problems in aging aircraft structures. After a certain period of service, exposure to corrosive environments, pitting corrosion, exfoliation corrosion, and stress corrosion cracking

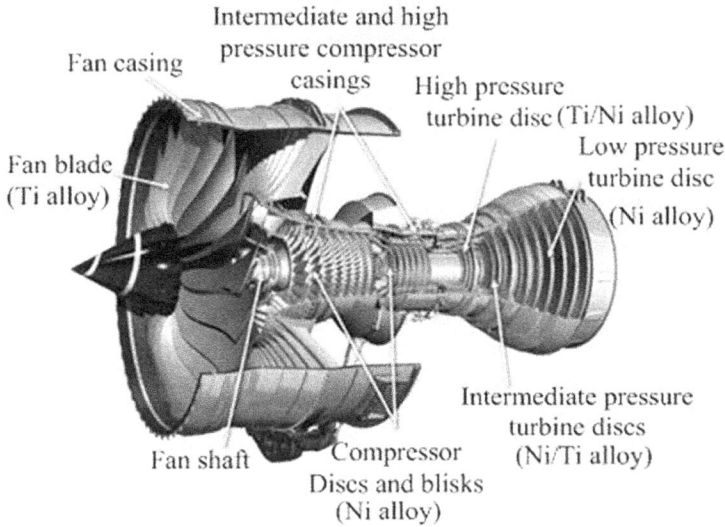

Intermediate and high
pressure compressor
casings

Fan casing

High pressure
turbine disc (Ti/Ni alloy)

Low pressure
turbine disc
(Ni alloy)

Fan blade
(Ti alloy)

Intermediate pressure
turbine discs
(Ni/Ti alloy)

Fan shaft   Compressor
Discs and blisks
(Ni alloy)

Figure 8.10 Various Ni- and Ti-based alloys used in the manufacturing of Rolls-Royce XWB Turbofan engine [39, 40].

may develop in aircraft components made of Al alloy. Claddings have the ability to deposit a different or similar metal or alloy onto a localized substrate surface. In general, the Al alloys are used for the application of aircraft structure structural integrity.

Superalloys are universally used in extreme conditions such as high-temperature oxidation, fatigue, creep, elevated temperatures, and wear. In recent times, the Ni-based alloys have replaced the Ti alloys in the manufacturing of aircraft gas turbine components. Figure 8.10 represents the various alloys used in the manufacturing of different components of aircraft engines [35]. During the combustion process, there is a high risk involved in the corrosion and wear of these superalloys at extremely higher temperatures. Commercially used powder alloys are (Inconel 100 and 718), Astroloy, Rene (41, 88DT and Rene 95), Nimonic (80A and 105), Waspaloy, Hastelloy (X and S), ATI-718 Plus, Udimet (500 and 700), N18, and AM1 [36, 37]. N18 and AM1 superalloys are used in modern Rafale aircraft engines. However, the novel yttria-stabilized-Zirconia (YSZ) coatings are widely adopted by the aircraft industry [35]. YSZ coatings are beneficial and light in weight; therefore, these coating improves fuel efficiency and reduce noise. However, YSZ coating can be improved by the addition of more ceramics [38]. In this context, thermal barrier coatings (TBCs) can be substituted by further thermal barrier claddings also due to the pore-free layers.

Engineering applications such as turbine engines are exposed to very high temperatures (greater than 1000°C). Therefore, these components, such as combustors, nozzle, fuel vaporizers, rotary turbine blades, afterburners, and others, have to operate at these high temperatures. Due to very high temperatures under highly corrosive impurities, these components are bound to experience defects like hot corrosion, erosion, creep, and fatigue. In order to overcome this limitation of high temperature, thermal barrier claddings are generally applied which enables these engines to operate with enhanced operating efficiency at high temperatures without affecting the base material temperature by utilizing a cooling system inside hot-section components [41]. The main purpose of these claddings is to provide the resistance against the oxidation in extreme temperature conditions such as in a boiler, an aircraft's or airplane's outer surface and its engine components, nuclear power plants, thermal power plants, and so on. Various necessary properties for better performance of TBCs are low thermal conductivity, low weight and their ability to sustain very high-stress variations caused due to continuous heating and cooling of components. Additionally, TBCs also need to have the ability to reflect most of the radiant heat originating from the hot gases and to protect the substrate material of components.

Generally, the structure of conventional thermal barrier coating involves a complex multilayer coating system comprised of a thermal insulating ceramic topcoat, an adhesive bond coat, and a thermally grown oxide (TGO). TBCs have a two-layered structure, which involves a two-layer ceramic coating system (YSZ) and a highly temperature-stable material on top [41]. Manufacturing the various types of thermal barrier materials on substrate materials is a complex mechanical, thermal, and chemical process. Plasma spraying, electron beam physical vapor deposition, air plasma spray, chemical vapor deposition, laser cladding, and microwave claddings are a few deposition techniques used for thermal barrier materials [42, 43]. The properties and life span of various TBC is strongly dependent on the microstructure of thermal-coated material [44]. The fine-grain structure of the TBC will cause strong adhesion between coating material and substrate, which will improve the performance of thermal barrier claddings [45].

The use of TBC as a topcoat for thermal protection results in enhanced life of components and enhanced overall efficiency. The various materials used as topcoat in TBCs are glass ceramic (ZnO-Al_2O_3-SiO_2, MgO-Al_2O_3-TiO_2), $La_2Zr_2O_7$, $Nd_2Zr_2O_7$, and others [46, 47]. Ceramic (yttria-stabilized zirconia, 7YSZ) is the most commonly used top layer material in TBC due to its high melting point (2700°C) and low thermal conductivity, among all ceramics [47]. Hexaaluminates, pyrochlores, perovskites, and zirconia doped with different rare earth (RE) cations (ZrO_2-Y_2O_3-Gd_2O_3-Yb_2O_3) are some alternative material groups that can be used as topcoat materials.

To protect the substrate from oxidation and to lessen the thermal expansion mismatch between the metallic substrate and the top coating layer, an intermediate layer of bond coat is also applied. NiCoCrAlY and CoNi-CrAlY are various materials used as bond coating in TBCs. A comparative study shows that of NiCoCrAlY bond coat utilized in TBC have much better characteristics for crack resistance and oxidation resistance than CoNi-CrAlY [48]. In modern times, research is being done on bond-less coatings to address the issues of uncontrolled growth of TGO layers as well as a reduction of TBC systems. A unique TBC system with only a topcoat layer and substrate material can be developed. Ni-based superalloy (NiCrAlY) processed with selective laser melting and reinforced with carbon nanotubes (CNTs) provides a possibility of eliminating bond coating. Hardness and elasticity of composites is enhanced by 2 times upon addition of only 1wt.% of CNT [49].

TGO is created along the coating/substrate interface during heat exposure, resulting in coalescence propagation and fracture formation along the topcoat/TGO interface. A large-scale buckling of TBCs would ultimately lead to the coating spallation that would eventually lead to the coating's collapse [50, 51]. Research is being carried out continuously in the field of TBCs so that conventional materials can be replaced with new ones that have less TGO generation, that is, enhanced performance of TBCs.

8.6 Conclusion

Microwave claddings are the modern approach for improving the surface properties of the different materials. Microwave energy utilization in surface coatings of substrate materials with cladding materials has the advantages of low power consumption, less processing time, less rapid heating, uniformity, and volumetric heating of substrate material. Various clad materials used in engineering applications are tungsten-based clads (WC-Co-Ni, WC10Co2Ni, WC-12Co), nickel-based clads (EWAC + 20% WC10Co2Ni, Ni-WC), Ni-SiC mixture, and cobalt-based alloys (EWAC-1006EE). In the presence of the acidic environment, the addition of hard ceramics and the creation of an oxide layer on the substrate surface are both responsible for materials exhibiting lower corrosion rates. In the case of an aqueous or a chloride environment, cladding material obstructs the movement of ions, leading to minimal material loss or decay. Corrosion proceeds at much faster rate in aqueous solutions than in acidic solutions. The use of a polymer cladding or coating over metals provides corrosion protection in two ways, that is, by anodic protection or by introduction of physical barrier between metal and environmental conditions. The cladding of polymeric compounds such as polypyrrole, polyaniline, polythiophene, and others protects substrate material with the help of anodic protection. On the other hand, microwave-assisted polyester cladding is used over steel metal to improve its corrosion resistance by creating physical barrier between the steel and its surroundings. This chapter provided insights to a microwave cladding-based metal deposit technique to articulate restoration of structural geometry affected by corrosion damage like fatigue (weakness in metal due to excess of use).

References

[1] J. Singh, S. Kumar, G. Singh, Taguchi ' s approach for optimization of tribo-resistance parameters forss304, *Mater. Today Proc.* 5 (2018) 5031–5038. https://doi.org/10.1016/j.matpr.2017.12.081.

[2] J. Singh, S. Kumar, S.K. Mohapatra, Tribological performance of Yttrium (III) and Zirconium (IV) ceramics reinforced WC—10Co4Cr cermet powder HVOF thermally sprayed on X2CrNiMo-17-12-2 steel, *Ceram. Int.* 45 (2019) 23126–23142. https://doi.org/10.1016/j.ceramint.2019.08.007.

[3] J. Singh, S. Singh, Neural network prediction of slurry erosion of heavy-duty pump impeller/casing materials 18Cr-8Ni, 16Cr-10Ni-2Mo, super duplex 24Cr-6Ni-3Mo-N, and grey cast iron, *Wear.* 476 (2021) 203741. https://doi.org/10.1016/j.wear.2021.203741.

[4] S. Singh, D. Gupta, V. Jain, Recent applications of microwaves in materials joining and surface coatings, *Proc. Inst. Mech. Eng. Part B J. Eng. Manuf.* 230 (2016) 603–617. https://doi.org/10.1177/0954405414560778.

[5] R.W. Bruce, A.W. Fliflet, H.E. Huey, C. Stephenson, M.A. Imam, Microwave sintering and melting of titanium powder for low-cost processing, *Key Eng. Mater.* 436 (2010) 131–140. https://doi.org/10.4028/www.scientific.net/KEM.436.131.

[6] K. Ishizaki, K. Nagata, T. Hayashi, Production of pig iron from magnetite ore-coal composite pellets by microwave heating, *ISIJ Int.* 46 (2006) 1403–1409. https://doi.org/10.2355/isijinternational.46.1403.

[7] T.J. Appleton, R.I. Colder, S.W. Kingman, I.S. Lowndes, A.G. Read, Microwave technology for energy-efficient processing of waste, *Appl. Energy.* 81 (2005) 85–113. https://doi.org/10.1016/j.apenergy.2004.07.002.

[8] R. Roy, D. Agrawal, J. Cheng, S. Gedevanlshvili, Full sintering of powdered-metal bodies in a microwave field, *Nature.* 399 (1999) 668–670. https://doi.org/10.1038/21390.

[9] D. Gupta, A.K. Sharma, Development and microstructural characterization of microwave cladding on austenitic stainless steel, *Surf. Coatings Technol.* 205 (2011) 5147–5155. https://doi.org/10.1016/j.surfcoat.2011.05.018.

[10] P.K. Loharkar, A. Ingle, S. Jhavar, Parametric review of microwave-based materials processing and its applications, *J. Mater. Res. Technol.* 8 (2019) 3306–3326. https://doi.org/10.1016/j.jmrt.2019.04.004.

[11] D. Gupta, A.K. Sharma, Investigation on sliding wear performance of WC10Co2Ni cladding developed through microwave irradiation, *Wear.* 271 (2011) 1642–1650. https://doi.org/10.1016/j.wear.2010.12.037.

[12] D. Gupta, A.K. Sharma, Microstructural characterization of cermet cladding developed through microwave irradiation, *J. Mater. Eng. Perform.* 21 (2012) 2165–2172. https://doi.org/10.1007/s11665-012-0142-2.

[13] D. Gupta, A.K. Sharma, Microwave cladding: A new approach in surface engineering, *J. Manuf. Process.* 16 (2014) 176–182. https://doi.org/10.1016/j.jmapro.2014.01.001.

[14] S. Zafar, A.K. Sharma, Development and characterisations of WC-12Co microwave clad, *Mater. Charact.* 96 (2014) 241–248. https://doi.org/10.1016/j.matchar.2014.08.015.

[15] S. Zafar, A.K. Sharma, Dry sliding wear performance of nanostructured WC-12Co deposited through microwave cladding, *Tribol. Int.* 91 (2015) 14–22. https://doi.org/10.1016/j.triboint.2015.06.023.

[16] A. Pathania, S. Singh, D. Gupta, V. Jain, Development and analysis of tribological behavior of microwave processed EWAC + 20% WC10Co2Ni composite cladding on mild steel substrate, *J. Manuf. Process.* 20 (2015) 79–87. https://doi.org/10.1016/j.jmapro.2015.09.007.

[17] R.B. Lohit, P.M. Bhovi, Development of Ni-WC composite clad using microwave energy, Mater. *Today Proc.* 4 (2017) 2975–2980. https://doi.org/10.1016/j.matpr.2017.02.179.

[18] A. Babu, H.S. Arora, S.N. Behera, M. Sharma, H.S. Grewal, Towards highly durable bimodal composite claddings using microwave processing, *Surf. Coatings Technol.* 349 (2018) 655–666. https://doi.org/10.1016/j.surfcoat.2018.06.059.

[19] A.M. Hebbale, M.S. Srinath, Taguchi analysis on erosive wear behavior of cobalt based microwave cladding on stainless steel AISI-420, *Meas. J. Int. Meas. Confed.* 99 (2017) 98–107. https://doi.org/10.1016/j.measurement.2016.12.024.

[20] I. Finne, Erosion of surfaces, *Wear.* 3 (1960) 87–103. https://doi.org/10.1016/0043-1648(60)90055-7.

[21] A. V. Levy, *Solid Particle Erosion and Erosion-Corrosion of Materials*, ASM International, Ohio, 1996.

[22] R. Bellman, A. Levy, Erosion mechanism in ductile metals, *Wear.* 70 (1981) 1–27. https://doi.org/10.1016/0043-1648(81)90268-4.

[23] Y.F. Wang, Z.G. Yang, Finite element model of erosive wear on ductile and brittle materials, *Wear.* 265 (2008) 871–878. https://doi.org/10.1016/j.wear.2008.01.014.

[24] J. Singh, Investigation on slurry erosion of different pumping materials and coatings, Thapar Institute of Engineering and Technology, Patiala, India, 2019.

[25] B. Singh, S. Zafar, Slurry erosion performance of Ni + xCr7C3 microwave composite clad with [10–30 wt%] Cr7C3 content, *Tribol. Trans.* 64 (2021) 528–539. https://doi.org/10.1080/10402004.2020.1863537.

[26] A. Babu, H.S. Arora, R.P. Singh, H.S. Grewal, Slurry erosion resistance of microwave derived Ni-SiC composite claddings, *Silicon.* (2021). https://doi.org/10.1007/s12633-020-00849-9.

[27] H.S. Grewal, R.B. Nair, H.S. Arora, Complex concentrated alloy bimodal composite claddings with enhanced cavitation erosion resistance, *Surf. Coatings Technol.* 392 (2020) 125751. https://doi.org/10.1016/j.surfcoat.2020.125751.

[28] J. Yan, N.M. Heckman, L. Velasco, A.M. Hodge, Improve sensitization and corrosion resistance of an Al-Mg alloy by optimization of grain boundaries, *Sci. Rep.* 6 (2016) 1–10. https://doi.org/10.1038/srep26870.

[29] M. Liu, Y. Guo, J. Wang, M. Yergin, Corrosion avoidance in lightweight materials for automotive applications, *Npj Mater. Degrad.* 2 (2018). https://doi.org/10.1038/s41529-018-0045-2.

[30] S.S. Rajahram, T.J. Harvey, R.J.K. Wood, Erosion-corrosion resistance of engineering materials in various test conditions, *Wear.* 267 (2009) 244–254. https://doi.org/10.1016/j.wear.2009.01.052.

[31] S. Kaushal, B. Singh, D. Gupta, H. Bhowmick, V. Jain, An approach for developing nickel—alumina powder-based metal matrix composite cladding on SS-304 substrate through microwave heating, *J. Compos. Mater.* 52 (2018) 2131–2138. https://doi.org/10.1177/0021998317740732.

[32] P.K. Loharkar, A. Ingle, Microwave-assisted corrosion resistant pure polyester coating of cold rolled closed annealed steel, *Mater. Today Proc.* 44 (2021) 1676–1680. https://doi.org/10.1016/j.matpr.2020.11.839.

[33] T. Ohtsuka, Corrosion protection of steels by conducting polymer coating, *Int. J. Corros.* 2012 (2012) 1–7. https://doi.org/10.1155/2012/915090.

[34] A.F. Baldissera, C.A. Ferreira, Coatings based on electronic conducting polymers for corrosion protection of metals, *Prog. Org. Coatings.* 75 (2012) 241–247. https://doi.org/10.1016/j.porgcoat.2012.05.004.

[35] A.P. Mouritz, *Introduction to Aerospace Materials*, 1st ed., Elsevier, Amsterdam, The Netherlands, 2012.

[36] M.L. Grilli, D. Valerini, A.E. Slobozeanu, B.O. Postolnyi, S. Balos, A. Rizzo, R.R. Piticescu, critical raw materials saving by protective coatings under extreme conditions : A review of last trends in alloys and coatings for aerospace engine applications, *Materials (Basel).* 14 (2021) 1656.

[37] R.L. Dreshfield, H.R. Gray, P/M super alloys: A troubled adolescent ?, *NASA Tech. Memo.* 8324. P/M 84 (1984) 23. https://core.ac.uk/download/pdf/42848789.pdf.

[38] A. Gloria, R. Montanari, M. Richetta, A. Varone, Alloys for aeronautic applications: State of the art and perspectives, *Metals (Basel).* 9 (2019) 1–26. https://doi.org/10.3390/met9060662.

[39] X. Liang, Z. Liu, B. Wang, State-of-the-art of surface integrity induced by tool wear effects in machining process of titanium and nickel alloys: A review, *Meas. J. Int. Meas. Confed.* 132 (2019) 150–181. https://doi.org/10.1016/j.measurement.2018.09.045.

[40] R. M'Saoubi, D. Axinte, S.L. Soo, C. Nobel, H. Attia, G. Kappmeyer, S. Engin, W.M. Sim, High performance cutting of advanced aerospace alloys and composite materials, *CIRP Ann.—Manuf. Technol.* 64 (2015) 557–580. https://doi.org/10.1016/j.cirp.2015.05.002.

[41] I.K. Igumenov, A.N. Aksenov, Thermal barrier coatings on gas turbine blades: Chemical vapor deposition (Review), *Therm. Eng.* 64 (2017) 865–873. https://doi.org/10.1134/S0040601517120035.

[42] S. Mahade, A. Venkat, N. Curry, M. Leitner, S. Joshi, Erosion performance of atmospheric plasma sprayed thermal barrier coatings with diverse porosity levels, *Coatings.* 11 (2021) 1–21. https://doi.org/10.3390/coatings11010086.

[43] J.R. Vargas Garcia, T. Goto, Thermal barrier coatings produced by chemical vapor deposition, *Sci. Technol. Adv. Mater.* 4 (2003) 397–402. https://doi.org/10.1016/S1468-6996(03)00048-2.

[44] D. Ye, W. Wang, H. Zhou, H. Fang, J. Huang, Y. Li, H. Gong, Z. Li, Characterization of thermal barrier coatings microstructural features using terahertz

spectroscopy, *Surf. Coatings Technol.* 394 (2020) 125836. https://doi.org/10.1016/j.surfcoat.2020.125836.

[45] K. Huang, W. Li, K. Pan, X. Lin, A. Wang, Microstructure and corrosion properties of La2Zr2O7/NiCoAlY thermal barrier coatings deposited on inconel 718 superalloy by laser cladding, *Coatings.* 11 (2021) 1–12. https://doi.org/10.3390/coatings11010101.

[46] T. Liu, X.T. Luo, X. Chen, G.J. Yang, C.X. Li, C.J. Li, Morphology and size evolution of interlamellar two-dimensional pores in plasma-sprayed La2Zr2O7 coatings during thermal exposure at 1300 °C, *J. Therm. Spray Technol.* 24 (2015) 739–748. https://doi.org/10.1007/s11666-015-0236-0.

[47] E. Bakan, R. Vaßen, Ceramic top coats of plasma-sprayed thermal barrier coatings: Materials, processes, and properties, *J. Therm. Spray Technol.* 26 (2017) 992–1010. https://doi.org/10.1007/s11666-017-0597-7.

[48] W.X. Weng, Y.M. Wang, Y.M. Liao, C.C. Li, Q. Li, Comparison of microstructural evolution and oxidation behaviour of NiCoCrAlY and CoNiCrAlY as bond coats used for thermal barrier coatings, Elsevier B.V, 2018. https://doi.org/10.1016/j.surfcoat.2018.08.024.

[49] S. Choudhary, A. Islam, B. Mukherjee, J. Richter, T. Arold, T. Niendorf, A. Kumar Keshri, Plasma sprayed Lanthanum zirconate coating over additively manufactured carbon nanotube reinforced Ni-based Composite: Unique performance of thermal barrier coating system without bondcoat, *Appl. Surf. Sci.* 550 (2021) 149397. https://doi.org/10.1016/j.apsusc.2021.149397.

[50] M. Karadge, X. Zhao, M. Preuss, P. Xiao, Microtexture of the thermally grown alumina in commercial thermal barrier coatings, *Scr. Mater.* 54 (2006) 639–644. https://doi.org/10.1016/j.scriptamat.2005.10.043.

[51] S.K. Essa, K. Chen, R. Liu, X. Wu, M.X. Yao, Failure mechanisms of APS-YSZ-CoNiCrAlY thermal barrier coating under isothermal oxidation and solid particle erosion, *J. Therm. Spray Technol.* 30 (2021) 424–441. https://doi.org/10.1007/s11666-020-01124-4.

Chapter 9

The Application of Microwave Energy for Fabrication of Polymer-Based Composite Materials

Manjeet Rani, Rajeev Kumar, Nishant Verma, Himanshu Pathak, and Sunny Zafar

Contents

9.1 Introduction

Fiber-reinforced polymer (FRP) composites have become widely used in aircraft and automotive industries. FRP composites have shown outstanding properties like higher fracture toughness, better damage tolerance, and high fatigue life. Carbon and glass fibers are primarily used fibers for critical structural applications like aerospace and wind turbine industries. In addition, polymer nanocomposites such as carbon nanotubes (CNTs)–based composites were developed to alter the material properties and added high values in different sectors like aerospace, energy, and health sectors [1, 2].

DOI: 10.1201/9781003248743-9

With the ever-increasing demand for FRPs, different curing methods have been developed to enhance the quality of mechanical, chemical, and physical properties. In addition, to reduce the overall cost of the product, various heating sources were used to shorten the production cycle time. For manufacturing sectors, the priority has been shifted toward the quality of products, lower cost, and short cycle time. In the competitive global market, cost reduction and short production cycle time are the priority of each manufacturing sector, especially in developed countries. So the manufacturing methods of composites play an essential role in enhancing the demand for composite materials worldwide. Manufacturing defects are critical factors that limit the usage of FRP from the desired or required specifications. The defects in the FRPs defined as irregularities that limit their application in a specific sector. Other types of defects like fiber misalignment cannot be avoided at the micro- and mesoscale. For decades, researchers have been trying to reduce the manufacturing defects for FRPs using different manufacturing techniques and alternation of processing parameters.

Damage and defects are different terms. Damage in FRPs can be seen only after the loading of FRPs in a specific application. Damage types are matrix cracking, fiber cracking, and delamination. Matrix cracks are induced due to thermal stresses and fiber cracks during the manufacturing of fibers. Manufacturing of FRPs without defects is highly expensive and not possible too. So researchers are focused on manufacturing the FRPs with minimum defects to minimize the manufacturing cost. In this point of view, characterization and quantification of defects are mandatory. So, according to the need and specific properties of the FRP composites, there is a need for an energy-efficient, sustainable, and fast heating-rate type curing heat source, that is, microwave energy.

Different curing phenomena of epoxy resin were observed during microwave and thermal heating sources. It was correlated to the different heating states of resin viscosity and temperature ramp rate (temperature profile) in the microwave and thermal heat. It is well understood that the change in temperature profile affects the crosslinking of epoxy resin. The percentage of crosslinking of epoxy is measured in terms of the degree of crosslinking and glass transition temperature of the cured composites. The degree of cure (crosslinking percentage), fiber–matrix interphase adhesion, and resin microstructure are the critical parameters to enhance the manufactured composite's mechanical performance.

Autoclave curing or conventional curing: fibers impregnated with resin are cured by applying heat and sufficient pressure to remove the air inside the composites. In this process, the composites are heated by conduction mode, and loss of energy occurs due to the heating of the autoclave chamber. The heating and pressurizing process occurred in the autoclave chamber. The composites manufactured by this process possessed high structural performance. The shortcomings of this process are high energy requirement and

long processing time that increases the overall production cost of the manufactured product.

9.1.1 Microwave curing

Microwave curing is an alternative technique to cure FRP composites. Microwave heating possesses characteristics like fast heating rate, energy-efficient,
uniform, and volumetric heating. It is a well-established fact that the approach
of microwave heating is distinct from the thermal heating approach. At
microwave exposure, microwaves interact with material molecules by polarization or reorientation in functional dipole groups. The reorientation in dipoles
restricts by some internal forces that result in the generation of kinetic energy.
The generated kinetic energy is converted into heat and results in the increment in temperature within the material. The generation of temperature
within the material enhances the uniformity and homogeneity of the heating
concept in the microwave heating approach [3–5]. In literature, this type of
heating is called an inside-out heating pattern that differs from the thermal
pattern, that is, outside in. Generally, for microwave curing of composites,
a curing cycle similar to the thermal cure cycle was adopted without any
modification or alternation in the process parameters (temperature ramp rate,
dwell temperature, change in viscosity of resin).

Tanrattanakal and Jaroendee [3] compared the mechanical performance
of the glass fiber-reinforced polymer (GFRP) composites cured with different heating sources, that is, microwave and conventional (thermal). The
effect of different step (one-, two-, and three-step) heating cycles with microwave power and the exposure time was also observed. The composites cured
with two- and three-step heating cycles possessed better mechanical performance than one-step and thermally cured laminates. This was correlated to
the sufficient temperature ramp rate at the beginning of the curing cycle and
accelerated adhesion between fiber and matrix material (epoxy resin). The
step heating is defined by the alternation in the power level (gradual increment) with time to alter the viscosity of the epoxy resin [3].

In another study, Li et al. [6] proposed a modified curing cycle to reduce
the residual strain, curing time, and energy consumption during the manufacturing of composites. This modified microwave curing cycle was used a
cyclic heating and cooling approach. In this process, the temperature fluctuation in the dwell cycle has been controlled to reduce the induced residual
stresses. The Fiber Bragg Grating (FBG) sensors are inserted within composite laminates to observe strains at the time of the curing process. Later,
the induced strain was converted into residual stress. In traditional heating,
the temperature fluctuation was controlled, but it shows limitations like
slow heating rate and increased production cost. With microwave heating,
it is possible due to a faster heating rate without additional production cost
[7]. Microwave curing of epoxy matrix material has feasible outcomes over

conventional curing, such as low energy consumption, moderate operating cost, less time to complete and have a uniform/consistent cure, enhanced mechanical/structural properties, and lower degradation [8]. Similar outcomes have been observed in previous studies during the microwave curing of FRP composite laminates [9–14].

To account for the importance of voids on mechanical performance/ properties of FRPs, extensive research was carried out and still going on. Some review articles were published with comprehensive details on the defects of the FRPs to provide cumulative insight. Processing (manufacturing process), structure (shape and size), and property (mechanical and physical) are these three parameters to define the voids in FRPs. The curing of the epoxy matrix in composite has a significant contribution in improving the structural and mechanical performance. The curing of epoxy depends on processing parameters like temperature, curing rate, and curing cycle time. The curing variation leads to voids, poor adhesion, and interfacial properties between fiber and matrix material [15].

This chapter aims to study microwave-cured composites for higher structural applications. It provides some insights regarding basic and innovative characterization techniques and testing methods to investigate the properties of manufactured composites. It also involves how the researchers optimized and found the properties to improve the quality of the composites that can be implemented for the critical application.

9.2 Microwave Interaction

9.2.1 Mechanism of Microwave Heating for Fiber-Reinforced Polymer Composites

Microwaves are electromagnetic radiations ranging from 300 MHz to 300 GHz and wavelengths from 1 mm to 1000 mm. These electromagnetic radiations have electric and magnetic components right-angled to each other. The electric component is responsible for heating nonmagnetic materials such as polymers and ceramics. In the case of magnetic materials such as iron, nickel, and so on, heat is generated due to the interaction of both electric and magnetic constituents with the material at the atomic level. The electric field contributes toward the heating of the material by dipolar loss and conduction loss. In dielectric materials, such as thermosetting polymers, dipoles are randomly oriented in different directions and try to align themselves in the direction of the applied electric field, which causes resistance and heat is generated in the material. In the case of conducting materials, such as copper, aluminum, carbon fiber, and others, free electrons start moving in the direction of electric field. The electro-magnetic force generated and the associated magnetic field oppose the movement of the electron in the direction of the applied field. The repeating oscillatory electric field produces heat in the material. Magnetic components resist the electron spin and domain orientation inside the material and cause heating [5].

In the case of the fiber-reinforced polymer composites (FRPCs), the microwave heating mechanism and dielectric properties of the polymer and fibers showed significant contributions individually. The dielectric constant and tangent loss of materials are the essential factors on which microwave heating depends during microwave exposure.

9.2.2 Microwave Interaction with Nonconducting Fiber-Reinforced Composites

The dielectric properties of nonconducting fibers such as glass, aramid, and natural fibers are close to thermoset resins. Therefore, fiber and resin are both equally responsible for heating the composites exposed under the microwave. The dielectric properties increase with a rise in the temperature of the composites and result in an increased heating rate. It resulted in more uniform curing of the composites, and the chances of hot spots can be minimized. Also, uniform heating produces superior interfacial properties between fiber and thermoset [14].

9.2.3 Microwave Interaction with Conduction Fiber-Reinforced Composites

The carbon fibers have high tangent loss values and absorb more microwave radiation and convert it into heat. The thermoset resin, such as epoxy, vinyl ester, and others, has significantly less interaction with microwaves due to their low dielectric properties. In manufacturing carbon fiber composites, a microwave interacts with the carbon fiber first, resulting in the heating

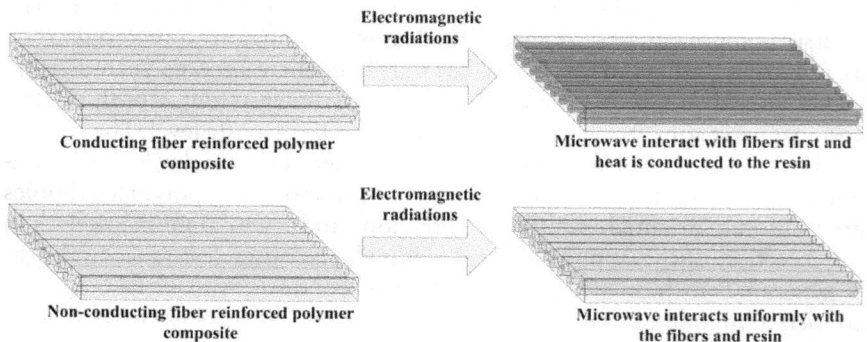

Electromagnetic radiations

Conducting fiber reinforced polymer composite

Microwave interact with fibers first and heat is conducted to the resin

Electromagnetic radiations

Non-conducting fiber reinforced polymer composite

Microwave interacts uniformly with the fibers and resin

Figure 9.1 Microwave interaction with fiber and matrix material: (a) Conducting fibers; (b) nonconducting fibers.

the of fibers, and then heat is transmitted to the resin in its vicinity through conduction mode. Thermosetting resin starts crosslinking during the curing process and causes a decrease in its dielectric values due to a change in viscosity. The difference in dielectric properties of resin can be attributed to the hindrance caused by the movement of the molecular dipole at high viscosity. The molecular dipole will not align freely along the electromagnetic field, making the material less responsive to the microwaves. Also, there are a few challenges in the microwave interaction of the conducting materials, such as arcing and hot spots. The arcing problem resulted from the low skin depth of conductors and reflected the microwave radiations [9]. Due to nonuniform heating, hot spots can degrade the composites at the surface of the composite. Many researchers used low microwave power levels to cope with these problems [16]. The interaction is shown in Figure 9.1.

9.2.4 Microwave Power and Microwave Exposure Time

The variation of microwave power and processing cycle time was studied to determine the maximum flexural strength and flexural modulus of glass fiber-reinforced polymer composites (GFRPCs). The microwave power, i.e., 400 W, 600 W, and 800 W, were used with a time range from 5 min to 20 min. At lower power, maximum cycle time was observed and vice versa. The maximum flexural strength was obtained with 800 W and 8 min (approximately). The maximum flexural strength was obtained with 600 W and a 20-min cycle time [17]. In later 1990s' studies, the presence of voids was observed more in the case of microwave cured composites than a thermal cure one. That was attributed to a short curing time and a lower pressure during the curing process. However, increment in transverse tensile strength and modulus was reported based on experimental results. The transverse tensile strength depends on the interfacial strength of the composite. Thus, fast curing rate with a low thermal gradient was the dominant factor for enhancing interfacial properties, not the short curing time [18]. The interfacial properties are essential for understanding mechanical properties and failure mechanisms of composites. A similar study was carried out to determine the influence of microwave and thermal curing on mechanical and interfacial properties of GFRPCs in 1995 by Yue and Looi [19]. Park et al. [20] investigated the mechanical performance (tensile strength, Young's modulus, flexural strength, modulus) of the microwave-cured carbon fiber composites and compared them with compression molding fabricated composites. In this study, microwave power and curing/processing time effect were selected as the processing parameters. Mechanical testing (tensile and three-point flexural tests) was carried out for the composites.

Less mechanical performance was observed at low microwave power (200 W) and 60-min curing time compared to conventionally cured samples. This could be correlated to insufficient curing of the resin; therefore, weak adhesion was obtained between the fiber mat and matrix material. Mechanical performance was enhanced at higher microwave power (400 W and 700 W) with the same curing time. Comparable mechanical properties were obtained with microwaves, and the thermally cured samples had significantly reduced curing times. The curing times were 60 min and 360 min for the microwave and conventional cured samples, respectively [20].

9.2.5 Voids Assessment

Voids are the most studied manufacturing defects. Voids assessment in manufactured composites is imperative to investigate the strength of the composite laminates quantitatively. Voids are considered matrix defects and come in the critical defects category that affects the mechanical performance of the composites. Voids are the spaces or regions that are unfilled with epoxy and fiber. Void formation and its effects on mechanical performance is still an active research area. The evaluation of voids' effect was started in the 1960s, and in the 1980s, researchers systematically started the voids formation analysis. The cross-sectional area images of the composite specimens can be used to estimate the voids. Scanning electron microscopic (SEM) or optical microscope (OM) images can quantify the voids or pores in composite specimens. The void ratio $\dfrac{total\,voids\,area}{total\,composite\,area}$ can be determined using the image filtering method.

Park et al. [20] observed the presence of voids in composites and established a correlation between delamination of composites and the presence of voids. The generation of epoxy grains was correlated with the degree of cure. If epoxy grains are generated, it acts as a driving force for delamination failure of composite laminates and simultaneously represents the presence of voids and an incomplete cure of epoxy. The generation of epoxy grains has been correlated with the curing cycle time and the microwave power [20]. When microwave power is insufficient and te curing cycle time is short, the generation of epoxy grains takes place. This indicates insufficient or incomplete curing of the composites. The phenomenon of incomplete curing enhances the delamination type of failure in composites. The presence of delamination failure and epoxy grains are shown in Figure 9.2.

9.2.6 Epoxy and Fiber Interface

Epoxy resins are widely accepted as matrix material in high and advanced structural applications due to their superior characteristics. Epoxy resins are chemically stable and have high mechanical properties. In past decades, epoxy resins were cured by conventional heating sources like autoclaves or conventional heating ovens. The drawbacks of conventional heating to cure the epoxy resins were identified: (1) consumption of a large amount of energy, (2) long processing time to cure the epoxy resins, and (3) large thermal gradient

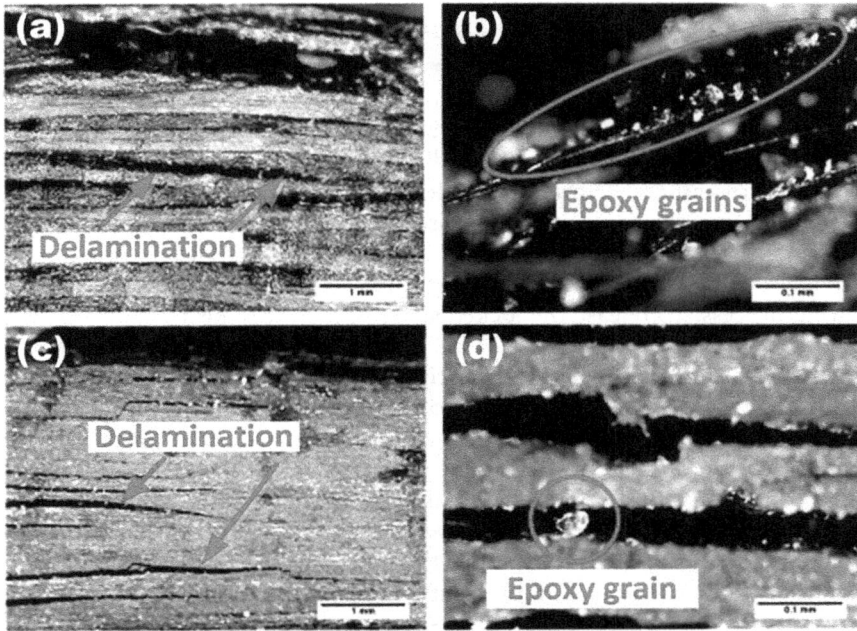

Figure 9.2 Optical microscopic images of the composites show the presence of delamination and grains of epoxy due to incomplete curing of composites.

Source: Reprinted with permission from Park et al. [20].

due to high heating rate. To encounter the reported drawbacks of conventional heating, microwaves as a heating source were proposed in recent studies. Microwave hearing is an innovative heating approach to cure epoxy-based composite materials. Microwave heating is innovative due to intrinsic features like uniform heating, controlled thermal gradient, avoiding a steep temperature rise, and no energy loss due to the volumetric heating approach.

When manufacturing fiber-reinforced polymer composites, there is a common problem in conventional and microwave heating approaches, that is, residual stresses. Residual stresses are induced in composites during manufacturing due to the different thermal behavior of the composite constituent materials. These residual stresses are the combination of normal, and shear stresses developed at the interfaces of the composites during the cyclic heating and cooling process. The change in the stresses during heating and cooling are shown in Figure 9.3. It has been presumed that there is no sliding on the interfaces during microwave heating due to volumetric heating phenomena. The constituent materials of the composites have different coefficients of thermal expansion (CTEs). During heating and cooling of the composites, the composite interface suffers stress due to different CTEs of constituent materials.

N. Li et el. [7] developed a semi-empirical theory to define the stress induced during the microwave curing process. At the start of curing of the epoxy, the loss modulus value of resin is infinity due to sol-gel liquid state. During microwave curing, the temperature of the resin starts to rise, and the sol-gel state of the resin is converted into glassy state due to crosslinking reaction mechanism. Before the gelation of the resin, the stored stresses can be relaxed, but after gelation, the resin start to restore the stresses owing to the viscoelastic properties of the sol-gelled rubber state. The crosslinking of the resin will lock the stresses in the composite until the completion of the curing process. To release the locked stresses, the cyclic heating and cooling method is proposed in this research. Owing to the volumetric heating characteristics of the microwave, fast heating of composite comes to the picture. With cyclic heating and cooling approach, the locked stresses in the heating cycle can be relaxed or released in the cooling cycle with same amount in the opposite direction. The stresses on the interface of the fiber and the resin and the shrinkage/expansion during heating/cooling can be account as applying loads on the resin matrix. With the rise or change in temperature during curing, the modulus of the fiber remains the same, but the elasticity modulus of the resin matrix changes with the temperature or curing process [7].

9.3 Characterization Techniques

9.3.1 Differential Scanning Calorimetry

The degree of cure or extent of crosslinking of epoxy resin was determined using the differential scanning calorimetry (DSC) analysis technique. In

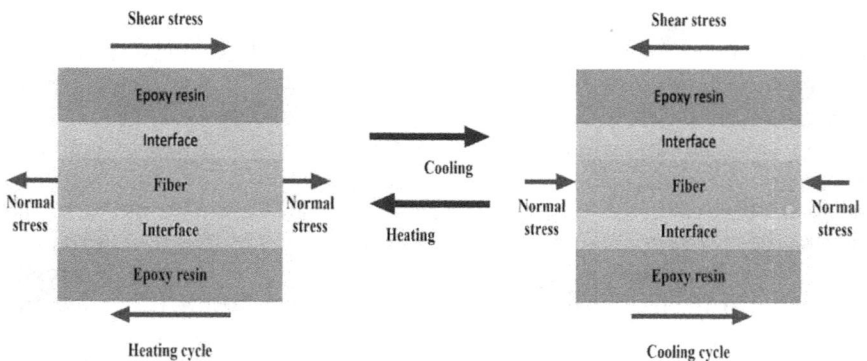

Figure 9.3 Schematic diagram of the fiber–matrix interface of the composite and change in the direction of stresses during heating and cooling cycles.

different studies, different heating rates and temperatures was used to investigate the degree of cure. The most used heating rate ranged from 8°C/min to 10°C/min [21–23]. Some studies showed a multistep heating phenomenon to cure composites using the microwave as a heating source [3, 24]. The obtained results demonstrated that multistep heating improves the mechanical performance of the composites compared to single-step or thermal heating. It was interesting to know the reasoning behind multistep heating or why it is better than thermal or single-step microwave heating approaches. Tanrattanakul and Jaroendee [3] determined the storage modulus for one-step and three-step microwave heating. The results showed a higher storage modulus for three-step heating (2×10^7 Pa) than single-step heating (1.2×10^7 Pa) as shown in Figure 9.4. DSC results showed different glass transition temperatures (T_g) (78 °C, 106 °C, and 115 °C) for one-, two-, and three-step, respectively, during the microwave heating process [3]. Generally, it was realized that the higher T_g of the composites showed higher strength and modulus. But up to some extent only, the mechanical properties rely on the T_g. In the case of carbon fiber–reinforced polymer composite laminates, thermally cured composites revealed lower T_g, but better mechanical performance than microwave cured composites. This was associated with the adhesion strength between fiber and matrix material [25].

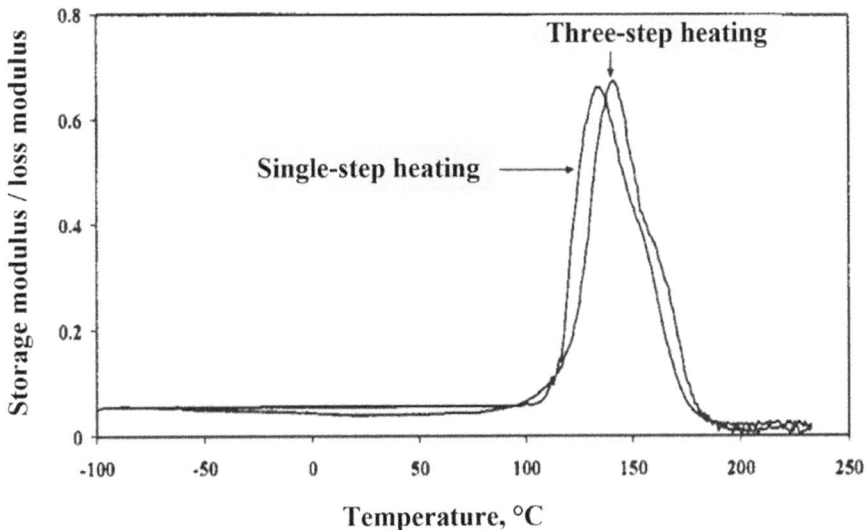

Figure 9.4 DSC graph shows higher storage modulus for three-step heating compared to single-step heating of composites cured by the microwave oven. Simultaneously, it shows a higher glass transition temperature for the three-step compared to the single-step heating process.

Source: This graph has been reproduced with permission from Tanrattanakul and Jaroendee [3].

9.3.2 Thermogravimetric Analysis

The thermogravimetric analysis (TGA) thermogram provides helpful information. TGA is generally used to find out the degradation behavior of the cured composites with the temperature variation. TGA is a technique in which the mass of a specimen in a specified atmosphere is observed against time and temperature. It measures the mass change in a specimen and recognizes oxidation, evaporation, and decomposition by changing temperature. One of the essential applications of TGA for the characterization of fiber-reinforced polymers is to determine the volume fraction of fibers in the composite. Maintaining the proper volume fraction of fibers is important for the quality control of composites to ensure their accurate mechanical and physical properties

9.3.3 Dynamic Mechanical Analysis

The dynamic mechanical analyzer testing setup can determine the dynamical thermal properties of the composite laminates. In previous studies, glass transition temperature, storage modulus (E'), and loss modulus (E'') of the composites were determined using dynamic mechanical analysis (DMA). The loss modulus shows the viscoelastic properties of composites during the curing cycle. During heating, the loss modulus increases, and at the gelation point, it shows maximum value. During cooling, the loss modulus goes down. With the help of DMA data, the degree of cure can be determined.

DMA is a highly adaptable and flexible analytical method to determine the physical properties of a wide range of materials. Until the 1950s, the commercial setup of DMA was not available to perform the testing. In the early 20th century, an initial attempt was started to perform various testing for composite materials. In the late 1980s, the computers were integrated with the DMA setup machines to make it easier for scientists and researchers. When it became popular among scientists, many suppliers contacted scientific laboratories and organizations for its commercial reach. Other techniques were added with the basic DMA like dynamic mechanical thermal analysis (DMTA), dynamic mechanical spectroscopy, or dynamic thermomechanical analysis. Now, DMA is well known for thermal analysis of the composites or a part of thermal analysis. Other thermal analysis methods are differential scanning calorimetry (DSC), thermo-gravimetric analysis (TGA), and thermo-mechanical analysis (TMA).

So, the DMA technique can be used to determine several mechanical and physical properties of a material, and its essential characteristic is to evaluate the T_g of a polymer and its composites. The DMA's sensitivity for T_g makes it an essential tool for researchers across the world. Despite T_g measurement, DMA can identify the secondary transitions during the thermal analysis of the materials. From this perspective, it is more beneficial for polymer-based composite materials. Although DMA has lots of benefits, it has some drawbacks also. For example, DMA is used to determine the storage modulus (E') of polymer-based material but to bring about an exact and precise value is very difficult, especially if the user is doing thermal scanning of the sample. With this variation, the mechanical performance of the material can differ from the expected results or performance. For near-accurate measurements of the material properties, the selected sample size and specification should match the standard sample specifications. For example, to determine the storage modulus of the FRP composites, the test should run isothermally, and sample specifications should be similar to standard sample specifications. Although it can sometimes be challenging to get accurate mechanical raw data using the DMA technique, the fundamental focus of the technique has always been to analyze a series of testing with the same sample geometry and testing conditions.

Dynamic mechanical analysis (DMA) is a versatile tool for determining the dynamic characteristics of materials. It can measure the properties of a range of materials, such as storage modulus (E', G'), loss modulus (E", G"), loss tangent (tan δ), glass transition temperature (T_g), and so on. The dependency of these properties on temperature can also be analyzed using DMA. The dynamic oscillating is applied to a specimen, and the material's response to the cyclic force is analyzed. The curves were plotted to identify transition temperature based on noticeable changes in the curve. DMA result of a CFRPC is shown in Figure 9.5. The variation in the value of storage modulus (E') is from 7 GPa to 30 GPa up to char temperature and reduced further. The viscoelastic material is given by loss modulus (E"). The value of E" suddenly drops with an increase in temperature. A peak of storage modulus was found at the gelation point (240°C). The value of the degree of cure at this point was about 0.78. The value of loss tangent decreased first up to gelation point and, after attaining a small peak, decreased further.

9.3.4 X-Ray Micro-Computed Tomography

Micro-computed tomography (micro-CT), a more recent innovation, permits many applications that include qualitative and quantitative analysis of composite laminates. This method/technique is based on observing the variation in attenuation along the path of an X-ray beam penetrated through a sample. This technique allows a cross-sectional image to be reconstructed mathematically, which is driven by the attenuation data collected at various angles using the detectors. The thickness of the two-dimensional (2D) image is known, and the image can be stacked virtually to create a 3D view from 2D images. The image's binary conversion (black-and-white) allows the voids to be segregated from the fiber and matrix material.

To investigate the voids/pores characteristics in FRPs, the X-ray micro-computed tomography technique is widely used. The voids characteristics

DMA Thermal Scan

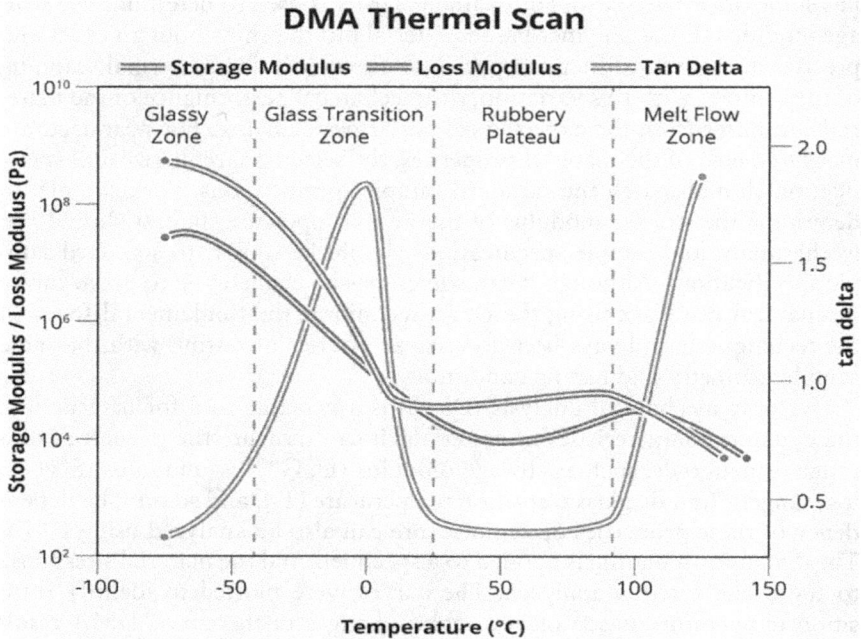

Figure 9.5 DMA thermal scanning of composite shows storage modulus, loss modulus, and loss tangent for a polymer-based composite material.

Source: Reprinted with permission from [7].

can be defined in terms of void content, location of voids, distribution frequency, shape (morphology), and size of voids. The micro-void percentage in transverse and axial flow for the same capillary number. The results showed higher micro-voids for the transverse flow of resin. In another study, micro-CT was used to determine the voids morphology (shape and size) and content in multidirectional composite laminates. The results showed fewer voids in a parallel direction than in the off-axis direction of fibers [26]. In carbon-fiber-reinforced polyether ether ketone (PEEK) composites, the voids were elongated (as shown in Figure 9.6) in the fiber direction observed by Comer et al. [27] through the micro-CT technique. In literature, composites were cured by laser and compared with thermally cured composites. The void content was found more in laser cured laminates, as shown in Figure 9.7c. The less void content in thermally cured composites was attributed to the prepreg roughness and processing parameters [28].

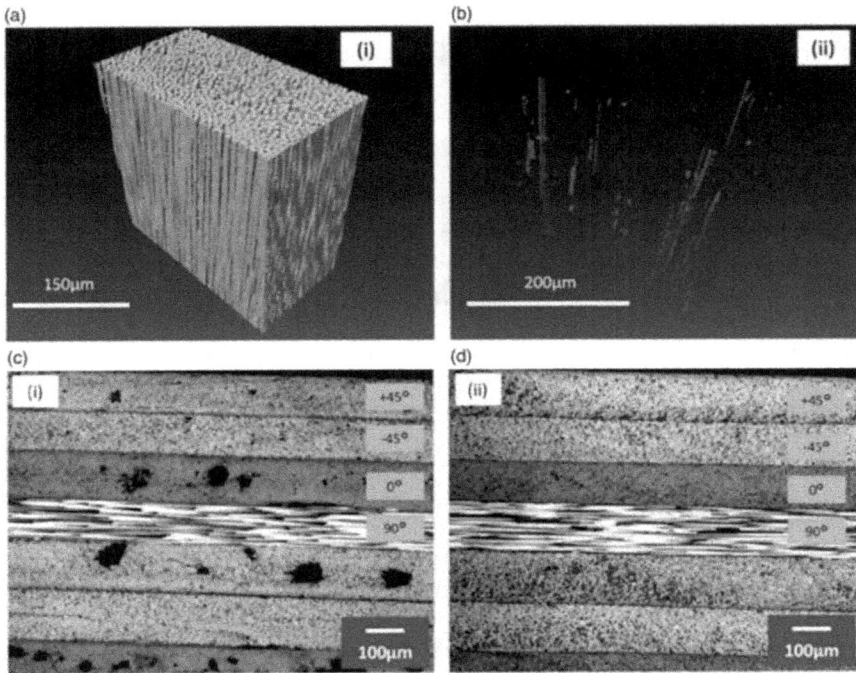

Figure 9.6 Micro-CT images of the carbon fiber–reinforced PEEK composites showing the presence of voids: (a) Carbon fibers, (b) the presence of voids in between fibers and matrix material; (c–d) voids' presence in composites cured with laser and thermal heating sources, respectively.

Source: Reprinted with permission from Mehdikhani et al. [26].

The samples of dimensions 230 mm × 10 mm × 2.6 mm, with voids introduced purposely and with optimized curing-cycle parameters, were manufactured [29]. The samples were scanned using X-ray CT (source voltage: 55 kV; number of projections: 2000; number of frames per projection: 4; source current: 140 mA) using a Nikon XTH320 system with the tungsten target material. The obtained data were used for post-processing within a large volume just inside the edges of the specimen and 50% segmentation threshold. This method enables an investigation of each void in terms of morphology, size, and location. For comparison purposes, high- magnification optical microscopy images and the micro-CT scans were used (as shown in Figure 9.7). The results demonstrate a good correlation in the void shape, dimensions, and distribution [30].

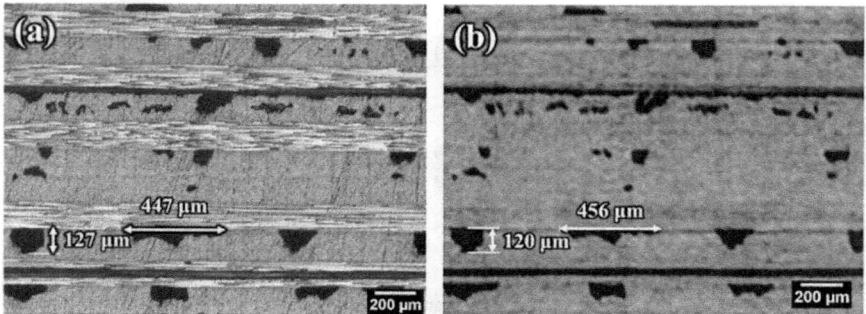

Figure 9.7 Micrograph images showing void size, shape, and distribution in composites scanned with (a) optical microscope and (b) micro-CT.

Source: Reprinted with permission from Tretiak and Smith [30].

9.4 Conclusion

The microwave processing of the materials in the future appears very demanding for specialized applications, and probably it will have limited usefulness for producing heat in conventional processes. In the domain of specialized applications, microwaves have distinct advantages as compared to conventional processing methods. Microwave processing applies to only certain materials, which can couple with electromagnetic radiations and produce heat. The lack of interaction among researchers, scientists, and industrialists results in the nonrealization of the fundamental importance of microwave processing. The primary type of microwave (generator, applicator, power supply) is available commercially. However, customizing the equipment in process design requires a unique applicator design to accommodate process variations.

In the case of the FRPCs, microwave processing depends upon the constituents' individual properties (matrix and reinforcement). The significant difference in the ability to absorb microwave radiation may result in nonuniform heating of the composite. Therefore, selecting fiber and matrix material is an essential criterion for microwave performance-enhancing the composite material's performance. Stacking sequence and fibers' orientation are the other essential factors to decide the ultimate properties of the composites. Thermoset material, such as epoxy has a certain gelation point, where crosslinking of the polymer starts. Therefore, it is essential to keep it under microwave exposure till the desired temperature reaches. A high degree of cure results in better interfacial bonding with the fibers and decreases porosity. There are certain kinds of fiber, which have high microwave-absorbing capacity. Under microwave exposure, these fibers interact more with the microwave, and more heat is generated. Heat transfer to low-temperature epoxy can result in a temperature gradient, which ultimately affects the mechanical performance of the composite laminates.

A proper combination of the fiber and matrix (having similar dielectric properties) may result in better properties of the composite materials. Intensive research on voids in FRP composites was carried out to observe the effects on mechanical properties. This is driven by industrial relevance on the effect of defects. The relevance of defects is enhanced by the rapid growth of composites in new and critical structural applications. More production by volume with quality requirements is needed to stay in competition worldwide.

Closed mold processes, such as vacuum-assisted resin-molding process, and compression molding, are known for high precision and repeatability. However, the cost may be the other factor to decide the final method to fabricate the FRPCs. Although many researchers have shown the benefits of microwave-assisted composites, much study and laboratory work are needed furthermore to shorten various challenges associated with FRPC fabrication.

References

[1] Jang JU, Park HC, Lee HS, Khil MS, Kim SY. Electrically and thermally conductive carbon fibre fabric reinforced polymer composites based on nanocarbons and an in-situ polymerizable cyclic oligoester. *Sci Rep* 2018;8:1–9. https://doi.org/10.1038/s41598-018-25965-w.

[2] Schwenke AM, Hoeppener S, Schubert US. Microwave synthesis of carbon nanofibers-the influence of MW irradiation power, time, and the amount of catalyst. *J Mater Chem A* 2015;3:23778–23787. https://doi.org/10.1039/c5ta06937h.

[3] Tanrattanakul V, Jaroendee D. Comparison between microwave and thermal curing of glass fiber-epoxy composites: Effect of microwave-heating cycle on mechanical properties. *J Appl Polym Sci* 2006;102:1059–1070. https://doi.org/10.1002/app.24245.

[4] Mgbemena CO, Li D, Lin M, Daniel P, Katnam KB, Kumar VT, خ.ال. Accelerated microwave curing of fibre-reinforced thermoset. *Compos Part A* 2018;115:88–103. https://doi.org/10.1016/j.compositesa.2018.09.012.

[5] Mishra RR, Sharma AK. Microwave-material interaction phenomena: Heating mechanisms, challenges and opportunities in material processing. *Compos Part A Appl Sci Manuf* 2016;81:78–97. https://doi.org/10.1016/j.compositesa.2015.10.035.

[6] Li N, Li Y, Wu X, Hao X. Tool-part interaction in composites microwave curing: Experimental investigation and analysis. *J Compos Mater* 2017;51:3719–3730. https://doi.org/10.1177/0021998317693674.

[7] Li N, Li Y, Jelonnek J, Link G, Gao J. A new process control method for microwave curing of carbon fibre reinforced composites in aerospace applications. *Compos Part B Eng* 2017;122:61–70. https://doi.org/10.1016/j.compositesb.2017.04.009.

[8] Chen M, Siochi EJ, Ward TC, McGrath JE. Basic ideas of microwave processing of polymers. *Polym Eng Sci* 1993;33:1092–1109. https://doi.org/10.1002/pen.760331703.

[9] Kwak M, Robinson P, Bismarck A, Wise R. Microwave curing of carbon-epoxy composites: Penetration depth and material characterisation. *Compos Part A Appl Sci Manuf* 2015;75:18–27. https://doi.org/10.1016/j.compositesa.2015.04.007.

[10] Wan J, Li C, Fan H, Li BG. Branched 1,6-Diaminohexane-derived aliphatic polyamine as curing agent for epoxy: Isothermal cure, network structure, and mechanical properties. *Ind Eng Chem Res* 2017;56:4938–4948. https://doi.org/10.1021/acs.iecr.7b00610.

[11] Li Y, Cheng L, Zhou J. Curing multidirectional carbon fiber reinforced polymer composites with indirect microwave heating. *Int J Adv Manuf Technol* 2018;97:1137–1147. https://doi.org/10.1007/s00170-018-1974-1.

[12] Hang X, Li Y, Hao X, Li N, Wen Y. Effects of temperature profiles of microwave curing processes on mechanical properties of carbon fibre-reinforced composites. *Proc Inst Mech Eng Part B J Eng Manuf* 2017;231:1332–1340. https://doi.org/10.1177/0954405415596142.

[13] Rao RMVGK, Rao S, Sridhara BK. Studies on tensile and interlaminar shear strength properties of thermally cured and microwave cured glass-epoxy composites. *J Reinf Plast Compos* 2006;25:783–795. https://doi.org/10.1177/0731684406063542.

[14] Zhou J, Li Y, Li N, Hao X, Liu C. Interfacial shear strength of microwave processed carbon fiber/epoxy composites characterized by an improved fiber-bundle pull-out test. *Compos Sci Technol* 2016;133:173–183. https://doi.org/10.1016/j.compscitech.2016.07.033.

[15] Mgbemena CO, Li D, Lin MF, Liddel PD, Katnam KB, Kumar VT, خ أو. Accelerated microwave curing of fibre-reinforced thermoset polymer composites for structural applications: A review of scientific challenges. *Compos Part A Appl Sci Manuf* 2018;115:88–103. https://doi.org/10.1016/j.compositesa.2018.09.012.

[16] Verma N, Kumar R, Zafar S, Pathak H. Vacuum-assisted microwave curing of epoxy/carbon fiber composite: An attempt for defect reduction in processing. *Manuf Lett* 2020;24:127–131. https://doi.org/10.1016/j.mfglet.2020.04.010.

[17] Boey FYC, Lee TH. Electromagnetic radiation curing of an epoxy/fibre glass reinforced composite. *Int J Radiat Appl Instrumentation Part* 1991;38:419–423. https://doi.org/10.1016/1359-0197(91)90118-L.

[18] Bai SL, Djafari V. Interfacial properties of microwave cured composites. *Composites* 1995;26:645–651. https://doi.org/10.1016/0010-4361(95)98913-6.

[19] Yue CY, Looi HC. Influence of thermal and microwave processing on the mechanical and interfacial properties of a glass/epoxy composite. *Composites* 1995;26:767–773.

[20] Park ET, Lee Y, Kim J, Kang BS, Song W. Experimental study on microwave-based curing process with thermal expansion pressure of PTFE for manufacturing carbon fiber/epoxy composites. *Materials (Basel)* 2019;12. https://doi.org/10.3390/ma12223737.

[21] Rangari VK, Bhuyan MS, Jeelani S. Microwave curing of CNFs/EPON-862 nanocomposites and their thermal and mechanical properties. *Compos Part A Appl Sci Manuf* 2011;42:849–858. https://doi.org/10.1016/j.compositesa.2011.03.014.

[22] Rojas JA, Ribeiro B, Rezende MC. Curing of glass fiber/epoxy resin composites using multiwalled carbon nanotubes buckypaper as a resistive element. *J Manuf Sci Eng* 2020;143:1–21. https://doi.org/10.1115/1.4048512.

[23] Li N, Li Y, Zhang L, Hao X. Kinetics modeling of carbon-fiber-reinforced bismaleimide composites under microwave and thermal curing. *J Appl Polym Sci* 2016;133:1–8. https://doi.org/10.1002/app.43770.

[24] Xu X, Wang X, Cai Q, Wang X, Wei R, Du S. Improvement of the compressive strength of carbon fiber/epoxy composites via microwave curing. *J Mater Sci Technol* 2016;32:226–232. https://doi.org/10.1016/j.jmst.2015.10.006.

[25] Fang X, Scola DA. Investigation of microwave energy to cure carbon fiber reinforced phenylethynyl-terminated polyimide composites, PETI-5/IM7. *J Polym Sci Part A Polym Chem* 1999;37:4616–4628. https://doi.org/10.1002/(SICI)1099-0518(19991215)37:24<4616::AID-POLA20>3.0.CO;2-W.

[26] Mehdikhani M, Gorbatikh L, Verpoest I, Lomov SV. Voids in fiber-reinforced polymer composites: A review on their formation, characteristics, and effects on mechanical performance. *J Compos Mater* 2019;53:1579–1669. https://doi.org/10.1177/0021998318772152.

[27] Comer AJ, Ray D, Obande WO, Jones D, Lyons J, Rosca I. Mechanical characterisation of carbon fibre-PEEK manufactured by laser-assisted automated-tape-placement and autoclave. *Compos Part A Appl Sci Manuf* 2015;69:10–20. https://doi.org/10.1016/j.compositesa.2014.10.003.

[28] Sisodia SM, Garcea SC, George AR, Fullwood DT, Spearing SM, Gamstedt EK. High-resolution computed tomography in resin infused woven carbon fibre composites with voids. *Compos Sci Technol* 2016;131:12–21. https://doi.org/10.1016/j.compscitech.2016.05.010.

[29] Gagauz I, Kawashita LF, Hallett SR. Effect of voids on interlaminar behaviour of carbon/epoxy composites. *ECCM 2016 — Proceeding 17th Eur Conf Compos Mater* 2016.

[30] Tretiak I, Smith RA. A parametric study of segmentation thresholds for X-ray CT porosity characterisation in composite materials. *Compos Part A Appl Sci Manuf* 2019;123:10–24. https://doi.org/10.1016/j.compositesa.2019.04.029.

Chapter 10

Microwave-Assisted Casting
A Key to the Metal Casting Industry

Gaurav Prashar, Hitesh Vasudev,
and Amit Bansal

Contents

10.1 Introduction

In the present scenario where the focus is on sustainable development; there is a need for such kinds of technologies to not only be economical but also efficient and reliable. Processing materials like metals, hard ceramics, polymers, and composites with better quality, shorter processing times, and, more important, having minimal influence on the environment is quite a challenging task. So there is a need of such kind of processing technologies that requires the least processing time with an eco-friendly nature. The unique characteristic features of material processing using the thermal energy of microwaves are getting attention from researchers, and more research is conducted joining/brazing, cladding, sintering, drilling, and metal casting with the assistance of microwaves [1–4]. Figure 10.1 shows the area of application of microwave technology. From the bar graph shown in Figure 10.2, it is found that the majority of microwave energy (ME) is required for the sintering application (58%), followed by cladding (20%), joining (15%), casting (4%), and drilling (3%), respectively [4].

DOI: 10.1201/9781003248743-10

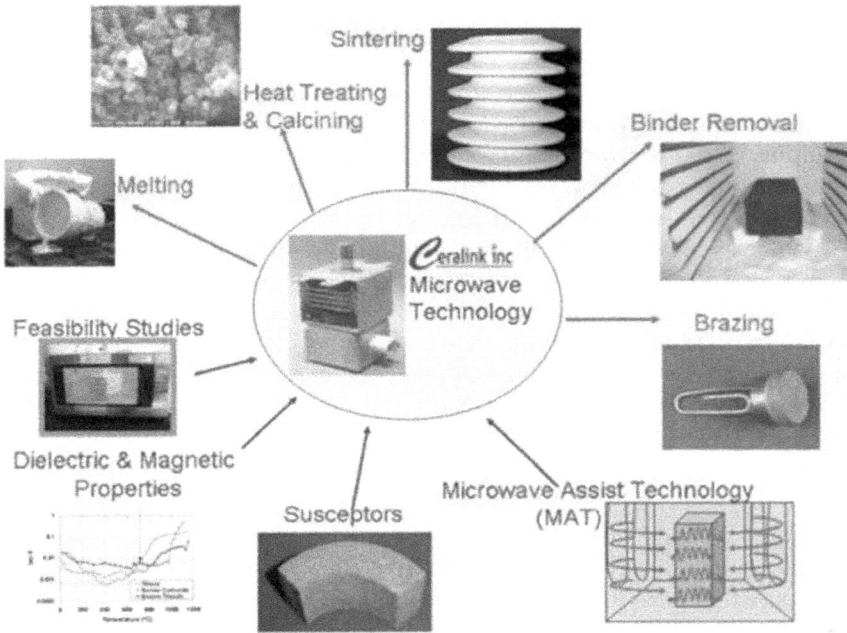

Figure 10.1 Broad areas of microwave technology.

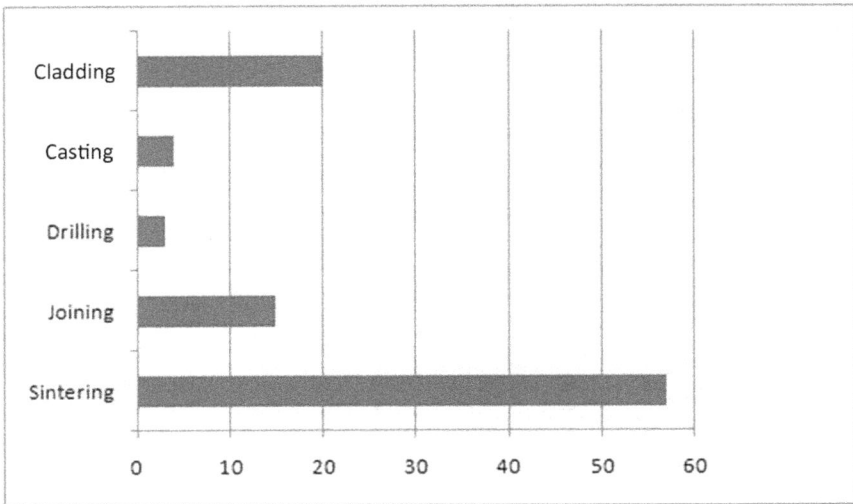

Figure 10.2 Microwave energy required for various methods.

During microwave irradiation, the electromagnetic energy is converted into heat energy depending on the polarizability of the material. In microwave heating, the heat is transferred at a molecular level. To get a good-quality casting, the design of tooling is a crucial step, and it should be selected based on material nature. The direct heating method is not suggested for metal-based systems due to microwaves' reflection from the bulk metals surface, resulting in the arc formation. Hence, for processing metals with ME, a novel method of microwave hybrid heating/indirect microwave exposure was discovered, and same must be implemented in different metal systems [3]. The feasible outcomes of direct and indirect microwave techniques are shown in Figure 10.3. The indirect microwave exposure used a suscep-tor material for absorbing the microwaves [5]. The direct heating method results in poor surface microstructures and mechanical properties because of fluctuating temperature gradient over the material [6]. A diagrammatic illustration of tooling needed for applications, such as casting, cladding, and heat treatment, is depicted in Figure 10.4. It comprises a susceptor and a masking material to safeguard the bulk metals against reflecting micro-waves. The mask shields the microwaves' direct contact with the metallic

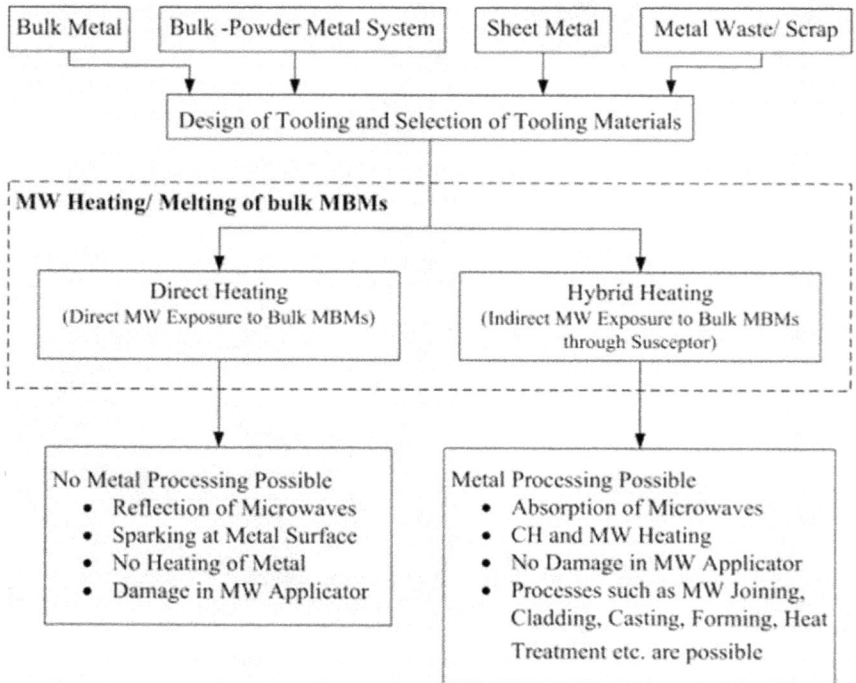

Figure 10.3 Procedure for microwave processing of metal systems [3].

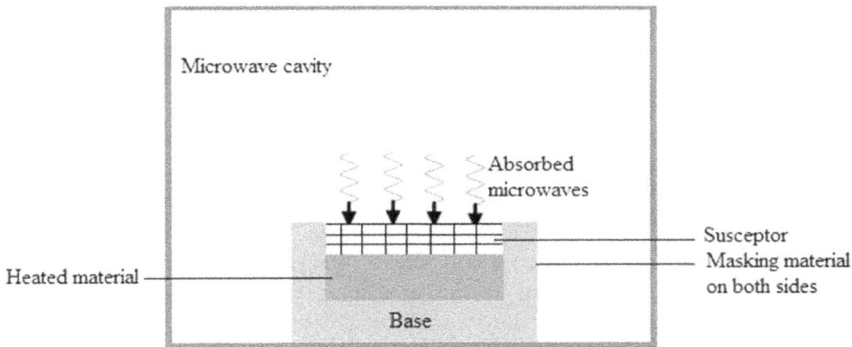

Figure 10.4 Microwave hybrid heating tooling setup.

material, whereas the function of the susceptor is to couple with the energy of microwaves and heat the metallic material from the outside via convection mode [6]. The susceptor converts ME into heat and functions as wireless heating element.

Microwave casting (MC) can be classified broadly into two categories: (a) in situ MC and (b) ex situ MC. In the in situ MC method, the melted charge passes through a sprue and finally enters the cope-and-drag section, where it solidifies and achieves desired shape, whereas, on other hand, by using MR, the charge is melted in the ex situ setup. The cope-and-drag section is placed outside the cavity [7].

10.2 History of Microwave Evolution

ME has been used in different applications over the last 65 years, including (a) low-temperature (<500°C) applications like synthesis and drying (between 1950–1970), (b) moderate-temperature range (500°C–1000°C) applications such as sintering (1970–1999), and (c) high-temperature (>1000°C) applications in processing advanced materials (1999–onward) [3]. Figure 10.5 depicts the significant historical advancements in the area of microwave processing technology. Metal probes are clearly playing a significant part in the processing of metallic-based materials, as evidenced by recent advancements.

Prior to 1999, it was a common misconception that metal-based materials reflect MR and that electron clouds form at sharp edges owing to MR's limited penetration into metallic materials, causing plasma generation and sparking. The foremost experimental work that reports on microwaves' interaction with metallic powders for increasing of heating rate by addition of a few percentages of electrically conducting metal powders was during refractory ceramics processing [8]. In the field of processing of metallic materials, limited research studies were available in sintering, brazing, and heating under certain conditions until 2008 [6, 9–12]. After 2008, intensive research studies were conducted for full sintering and heating of metal powder using microwaves, which leads to an enhancement in both the metallurgical and mechanical properties [13–18]. The processing of bulk metal using MR was reported first in the form of an Indian patent for joining of bulk metal in 2009 [19] and for cladding in 2010 [20] of various metallic and nonmetallic powders onto metal substrates.

In the area of metal casting by using ME, a US patent was filed in 2006 claiming the technique and apparatus for melting metallic metals [21]. Researchers used ceramic crucible for metal melting. The crucible was enclosed in a ceramic casket for proper insulation. An oxygen-free environment was used for metal heating and fully molten metal was poured into a mold placed outside or in a mold located just underneath the crucible. In another experimental setup, a technique of modeling and simulation was put into use for industrial microwaves to melt metals, and thereafter, the temperature profile outcomes were validated [22]. Wiedenmann et al. [23], in another effort, used a mode stirrer for even distribution of ME in the cavity of the applicator. The computer aided design (CAD) model was developed first for experimental setup, and then both experimental and simulated

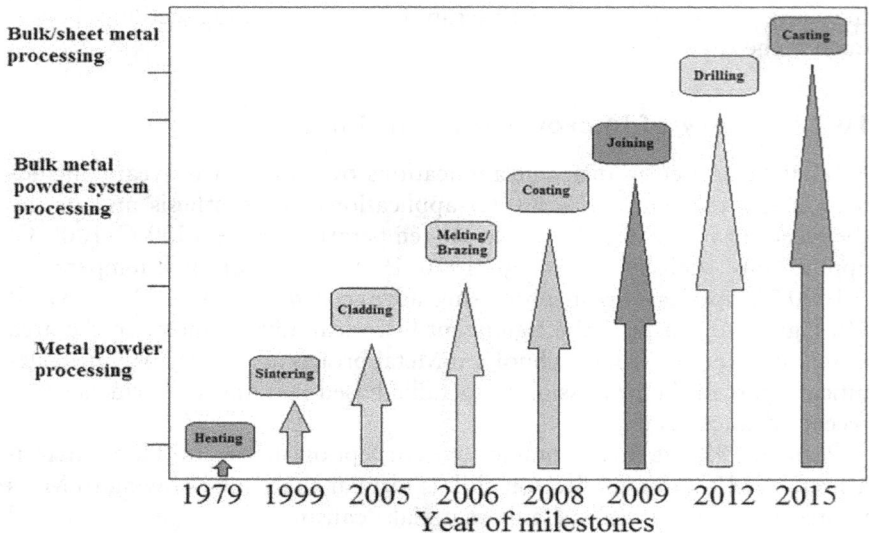

Figure 10.5 Microwave evolution history.

outcomes for the profiles of temperatures near to the mold were compared, after switching off the power of microwave. The technique and mold for thin-metal part castings using ME was reported in the patent form to minimize or eliminate casting defects like porosity, cold shunts, and incomplete extensions [24–25]. The published experimental literature in processing of metallic materials using microwave energy has highlighted the potential of microwaves but with proper tooling. Therefore, this opens up the opportunity windows for the technique of MC. Next section focuses on the experimental studies related to metal casting using ME.

10.3 Some Studies Related to Metal Casting Using Microwave Energy

An in situ MC method was developed with help of a domestic microwave applicator at power 900W and a frequency of 2.45 GHz. The MC of lead (Pb) was formed with help of a split mold (Figure 10.6a), having the cope-and-drag and the final developed Pb cast, and is depicted in Figure 10.6b. The in situ MC modeling process was conducted, and the simulated outcomes for SiC, Al, and graphite mold materials were reported [26]. From the results of the simulation, it was observed that both thermal and electromagnetic properties of mold material affect the in situ metal-cast properties. Furthermore, MC of bulk metallic materials was also explored in literature. The necessity of high specific strength in developed MC requires Al alloys processing by using MR. In a study, MC of the pure bulk Al alloy grade 1050 was reported using MR [27]. The enhancement in mechanical properties of the MC Al alloy was found when compared with the as-received Al alloy. The grain structure of as-cast Al alloy was observed to be uniform with a dominant presence of the Al phase when put under scanning electron microscopy (SEM) and X-ray diffraction (XRD) analysis. In another effort, in situ MC of Al alloy grade 7039 [28] was examined and it was reported that cast 7039 Al alloy has a porosity less than 2%. Although the castability of the Al alloy of grade 7039 was poor but due to its better strength and good toughness it founds its application in armor suits and military vessels. Finer microstructure with enhanced mechanical properties was noticed in the developed in situ microwave casts of Al Alloy 7039 [29]. Methods to refine the grain structure of microwave casts and the effect of grain morphology and intermetallic precipitates onto the indentation hardness during these methods were also examined [29–31]. Melting Al alloy having a grade 7039 in a room temperature was examined during the in situ MC method, which reveals that oxide layer presence plays a crucial role during in situ MC process [28].

Microwave heating of Cu was also studied in an argon (Ar) environment [32]. However, to explore the in situ MC of bulk Cu, Mishra and Sharma [33] developed Cu casts at a frequency of 2.45 GHz and a power of 1400W using graphite mold. The effect of the oxide layer during Cu melting was explored. The developed Cu casts were characterized using the SEM and XRD methods to examine the microstructure and mechanical properties. The authors reported that Cu casts were dense and homogeneous with porosity in range 2–5%, which was a matter of concern. Typical SEM micrographs of developed Cu casts are depicted in Figure 10.7, which indicates the potential of MC process. Marahadige et al. [34] cast naval brass material using a microwave oven (domestic). From the higher magnification SEM images of cast naval brass, one can see the nodular microstructure. The average value of ultimate tensile strength of developed casts was recorded to be 347.2 MPa.

MC of nickel-based ductile powder matrix with 10wt% SiC composite was observed to give better reinforcement distribution having equiaxed

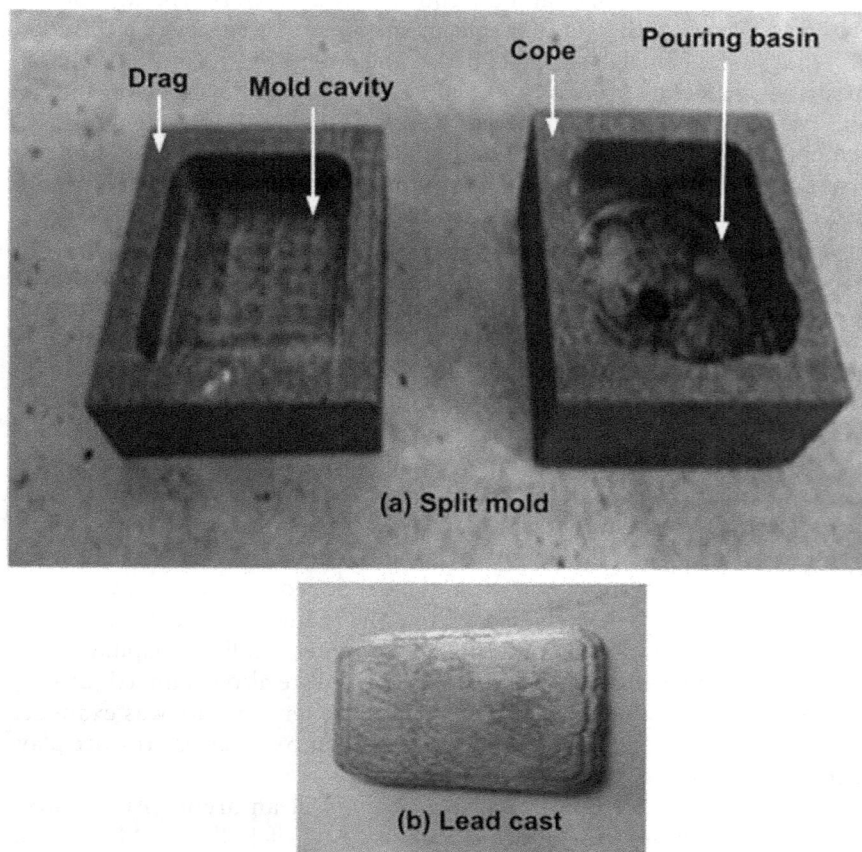

Figure 10.6 MC of lead (Pb): (a) cope and drag; (b) Pb cast [3].

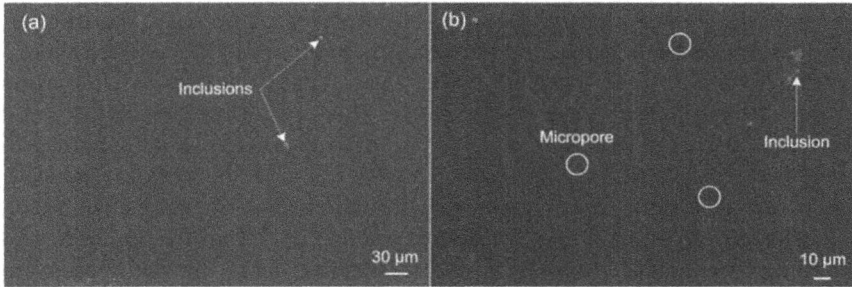

Figure 10.7 SEM micrographs of in situ cast [33].

Table 10.1 MC of Various Materials with Their Specifications [4]

Material/Reference	Casting Type	Type of Microwave	Power/ Frequency Range	Time of Exposure (s)
Al alloy 1050/ [27]	Ex situ	Domestic	900W/2.45GHz	1200
Al alloy 7039/ [28]	In situ	Industrial	1400W/2.45GHz	930
Al alloy 7039/ [31]	In situ	Industrial	1400W/2.45GHz	400
Al alloy 7039/ [38]	In situ	Industrial	1400W/2.45GHz	525
Al alloy 7039/ [30]	In situ	Industrial	1400W/2.45GHz	–
Al alloy 7039/ [7]	In situ/ ex situ	Industrial	1400W/2.45GHz	–
EWAC+10wt%SiC/ [36]	In situ	Domestic	900W/2.45GHz	108–1440
EWAC+ 5& 10wt%SiC/ [35]	In situ	Domestic	900W/2.45GHz	108–1440
EWAC+10wt%SiC/ [39]	In situ	Domestic	900W/2.45GHz	1320±180
EWAC+ 5& 10wt%WC-8Co/ [37]	In situ	Domestic	900W/2.45GHz	1440
Naval brass/ [34]	–	Industrial	3300W/2.45GHz	1020
Cu/ [33]	In situ	Industrial	1400W/2.45GHz	890±10

grains [35]. The micro-hardness value of developed composite casts was compared with that of conventional casting method. The authors found that the hardness of the casted specimen was higher and the porosity was less when compared with the conventional casting samples [36]. In another study related to composite coatings, Singh et al. [37] mixed WC-8Co in a ductile Ni matrix. The MC was carried out using the microwave oven having the frequency set at 2.45 GHz, the power at 900 W, and the exposure time to 25 min. From XRD analysis, the hard phases such as NiSi and $Cr_{23}C_6$ were revealed. The particles of WC were randomly dispersed into the matrix, which results in the development of intermetallic carbides leading to increase microhardness (788 ± 52HV) [37]. The various materials that were cast using ME are summarized in Table 10.1.

10.4 Characterization Techniques in MC

To know the changes taking place in metallurgical and mechanical properties for both as-received and -casted materials undergo requires various standard characterization techniques. Table 10.2 summarizes the various techniques followed by different researchers during their experimental studies, and there are still a few characterizations required that need to be explored to develop this MC technique as a future-ready commercialized industrial manufacturing technique. To examine the transformation of phases, the specimen should undergo X-ray diffraction under defined scanning angle and scanning step. Surface topography, surface morphology, elemental analysis, and failure mode should studied using SEM/EDAX method. Digital Vickers microhardness tester and tensile tests results should be put in for calculating microhardness and tensile strength of microwave-assisted casted specimens. The curve of time versus temperature describes the various steps involved in the MC process, like the absorption rate of MR and oxide layer formation.

10.5 Metal Melting Using ME (Case Study)

Traditionally, researchers thought that metals and microwaves have not been compatible with each other due to the presence of electron clod in the metallic materials, which causes reflection of microwaves with the bulk metallic pieces. Later, it has been reported that this assumption was valid only at room temperature. The metal in powder form or bulk metal can be processed at high temperatures by utilizing this ME in the form of hybrid microwave heating. Therefore, microwaves have remarkable possibilities for use in the field of metal melting. Ceralink industries developed and designed a technique to melt scrap Al using ME in a furnace designed with an embedded layer of SiC-susceptor blocks (Research Microwave Systems, Troy, NY). The SiC is heated in the range of 100–150°C/min, yielding a quick, clean Al melt (Figure 10.8). Researchers reported that there are remarkable savings in energy (0.7 kWh/kg consumed) in comparison with that consumed utilizing conventional reverbatory melting furnace. The yield rises from 94% to 99% via the elimination of the natural-gas environment. Furthermore, the MC method is green and is ten times quicker than conventional metal melting. Al charge (36 kg or 80 lb) was first melted and then cast only in 40 min. The microwave furnace for scaling up in the future is available with one of Ceralink's industrial partners. Different metals and alloys can be melted at a faster pace using microwave hybrid heating, like Cu, bronze, brass, gold, Au, tin, Fe, and Ti.

10.6 Role of Process Parameters in MC

The capacity of microwaves to heat materials is an important breakthrough in the field of science. The thermic potential of microwave has resulted in

Table 10.2 Characterization Involved in MC Materials [4].

Material/Reference	SEM/EDAX	XRD	Micro-hardness	Micro tensile Test	Time vs. Temperature Test	3-Point Bend Test	Factography	Porosity	Taguchi ANOVA Array	Fatigue Life
Al 1050 [27]	✓	✓	✓	✓			✓			
Al 7039 [28]	✓	✓	✓		✓			✓		
EWAC+10wt%SiC [36]	✓	✓	✓							
EWAC+ 5&10wt%SiC [35]	✓	✓	✓							
EWAC+10wt%SiC [39]	✓	✓	✓	✓						
EWAC+ 5&10wt%WC-8Co [37]	✓	✓	✓							
Naval brass [34]	✓	✓	✓	✓	✓		✓			
Cu [33]	✓	✓	✓		✓			✓		

Figure 10.8 (a) Microwave susceptor-assisted Al melting; molten Al flows out of the melting microwave furnace; (b) microwave-heated molten Al is cast into ingots; 80 lb of Al is melted in 40 minutes.

the innovation of appliances and techniques that could well replace in use conventional thermal techniques and minimize humankind's dependence on fossil fuels. Microwave heating in its beginning days was restricted only to drying and processing of food only. But researchers/professionals over the past decade have explored the usefulness of microwaves in processing materials with encouraging results. At present, research on materials processing using ME is focused on investigating the processing of new materials and accomplishing greater reliability and process control in the current processes. All this necessitates a thorough understanding of the process parameters used in the microwave heating method. The final quality of the MC relies directly on the microwave process parameters opted during the casting process. The microwave process parameters selected are the researcher's initial inputs (I/P) that have a significant impact on the ultimate results. Four major I/P parameters have been recognized by the researchers that governs the entire microwave-based heating. These I/P parameters are (a) magnetic and dielectric properties, (b) load, (c) applicators, and (d) heating mechanisms. All these I/P parameters are shown in Figure 10.9. Each of these I/P parameters is influenced directly by various factors like materials morphology and size being processed, selected frequency, applicator type and inside positioning, the setups employed, and the major heating mechanisms. The type of applicator (single or multimode), the material size, the heating mechanism, and the process temperature all require more attention and control. The solidification and melting rate of metals during MC influences both mechanical and metallurgical properties of the cast [40–41]. The alloy's solidification happens over a range of temperatures and comprises both solid and liquid phases. The solid phase during solidification nucleates first and then forms a dendritic structure. As a result, throughout the alloy

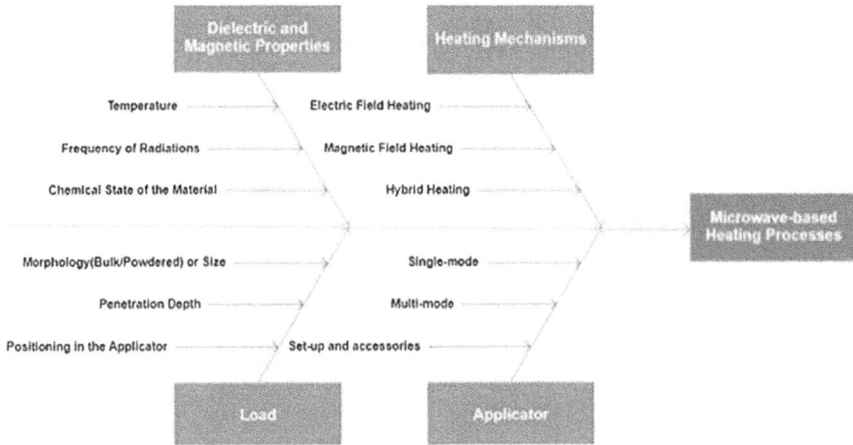

Figure 10.9 Various process parameters used in MC [43].

casting process, establish a solidification benchmark that improves the alloy cast's thermo-mechanical qualities [42].

The metallic materials heating owing to the MC process relies on the effective conversion of MR electromagnetic energy into heat energy that melts metal. This energy conversion via which bulk metallic materials melt fully relies on the heating characteristics and mechanical properties of mold setup [44].

Although in literature studies on melting of Al, Cu and steel samples using ME have been carried out by Agrawal [45], but if we talk about in relation to the experimental parameters, only limited studies exist. Chandrasekaran et al. [32] conducted similar studies to figure out the heating attributes of metals, like Pb, Sn, Al, and Cu using a multimode oven. Microwave power and the amount of material to be processed were chosen as process parameters, and experimental outcomes were compared straight away with the conventional melting approach. The authors found that microwave heating leads to the quick rise in temperature related to the enhanced interaction of ME with the experimental material. For the experiment, three distinct power levels: 100% (1300 W), 70% (910 W), and 40% (520 W) were used, as well as varying sample loads (25–150 g). A major outcome revealed by the authors that the melting of experimental materials via the power of microwaves was observed to be two times quicker than conventional heating-based melting, and it is also a cleaner process with reasonable process control.

The most common response parameter in all microwave heating areas is the rate with which temperature rise that indirectly impacts product qualities.

Control over this rate gives merits via enhancing the processed material properties and morphology.

Xu et al. [46] carried out the Cu powder melting by varying three powder sizes, namely, 25 μm, 38 μm, and 74 μm. In addition to this, the powder quantity was also varied, namely, 50 g, 100 g, and 150 g. The specimens were exposed between the power range 1.3–1.8 kW at a frequency of 2.45 GHz in an industrial applicator (multimode type). The applicator used had an arrangement for measuring temperature. The experiment conducted leads to the formation of a model for temperature rise in Cu of powder size 75 μm. This temperature rise has been expressed directly as a function of 't'. Authors also recommended that rise in temperature results in densification of castings from 2.13 gram/cm^{-3} at ambient temperature to 8.07 gram/cm^{-3} at 1090°C for the 74-μm size powder. Hence, the required cast quality level can be achieved by governing the temperature rise.

10.7 Future Directions

The prime issue associated with MC is its commercialization in the coming future which is still lacking when compared with the application area of conventional casting technology. The MC-through-ME route is a recent outcome/development for the casting industry, and hence, further deep research is required to establish its viability to put it on the commercial industry shop-floor level for casting different materials. The qualities that make MC a lucrative research area in future are the energy savings achieved with this technology and superior casting quality in relation with better mechanical properties achieved with better process control. The developments in the area of MC of metals started in 2005; limited research has been explored for the melting and solidification of both metals and metallic alloys. Subsequently, the technique and tooling for MC were designed. MC has gained researchers' attention owing to the implicit processing challenges associated with the conventional casting method, whereas less processing time coupled with remarkable cost savings make MC a better choice. A variety of well-advanced and special casting techniques in practical has defects, such as coarse grains, porosity, segregation, shrinkage cavity, and bottom shrinkage, which reduces plasticity and strength in developed casts [47]. Hence, techniques like die-casting, spraying forming, squeeze casting, and semi-solid forming, among others, were designed to reduce these casting problems [48–51]. These techniques homogenize the entire mechanical properties and enhanced the operating life of developed casts by minimizing segregation, improving, and refining the casting microstructures effectively. But nondendritic structure formed in these casts during casting operation requires further studies [52–53]. The use of ME can be the solution to control the nondendritic structure owing to its unique different heating properties. However, the MC of metallic-based materials using ME is a challenging task due to microwave reflection during interaction with metallic materials. However, recent developments in the tooling and methods for MC have exposed number of opportunities in this field. A possible research direction in MC is shown in Figure 10.10. The figure depicts the distinct dimensions and research opportunities that exist in the microwave field. Principally, the good outcome from a process relies on the clarification of the process physics. Unfortunately, a thorough understanding of the physics related with the melting of the metallic-based materials with ME is yet to be achieved. The research community is putting efforts to throw light onto the problem. It is, therefore, crucial to explore theoretical aspects first before divulging into the applications.

The process's physics mainly govern three main areas: (a) process development, (b) control of the process, and (c) quality control of MC. The process development relies on the knowledge gained on the interaction of microwave

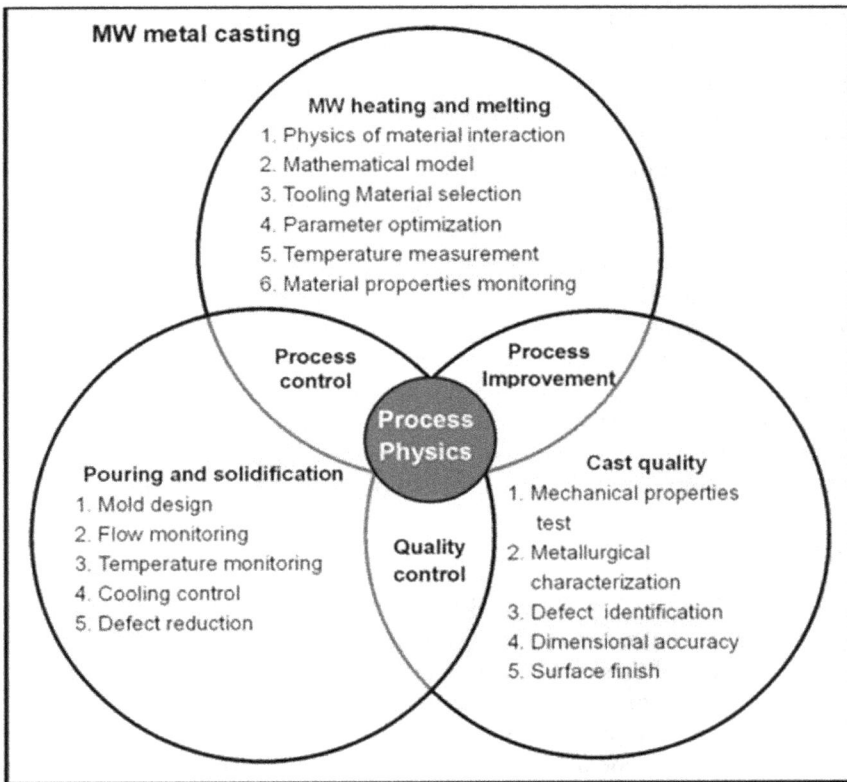

Figure 10.10 Future research directions in MC of metallic materials [3].

material and consequent heat mechanism. Control of the process is possible via control of microwave heating and melting, followed by the pouring and solidification of the melted material. The equipment design affects the microwave heating and melting aspects. The design of equipment needed knowledge of microwave-material interaction mechanism, requirement of power, operation frequency, time of exposure, and tooling materials (mold, susceptor, separator), and so on. Pouring of molten metal and its solidification rate are governed by the mold design during MC. Furthermore, the design of the mold was based on certain parameters like (a) fluid-flow physical aspects inside a mold, (b) the transfer of heat from the surge and from the mold to nearby surrounding environment, (c) the cooling effect on structure of the grains, (d) the friction behavior of mold material, (e) mold thickness, and (f) cooling conditions in microwave applicator. The quality control of MC can be governed via monitoring the pouring process, the solidification rate, and appropriate design of mold. The control of characteristics such as molten metal flow rate, the temperature of pouring, time, a loss of heat, and so on gives a better idea of the physics associated with pouring and solidification. The identification of a defect in MC components by different techniques and the calculations of the metallurgical and mechanical properties assist in quality control of developed casts. The enhancements needed in the MC method need the regular evaluation of casted components quality using standard equipment and then subsequently doing alterations in microwave heating and melting during casting process. The alterations in heating and melting during MC need in-process temperature monitoring and metal properties during the processing of metallic-based materials.

The method of heating and melting through ME can be optimized for good product quality via optimizing microwave parameters setting such as the power range, the time of exposure, the environment of the applicator, and the cooling gas flow. Thus, deep understanding and knowledge of process physics results in good design of equipment which gives better solidification method of melted metal in mold, which results in better cast quality. Thus, addressing these design problems would result in better MC technology and same to be adopted in near future as a commercialized solution.

10.8 Conclusion

This review chapter has discussed the potential of microwaves in processing of metal-based materials. MC is divided mainly into two major groups, that is, in situ and ex situ MC. The microwave processing of Al alloys, Cu, bronze, lead, and composites has gained attraction because of merits associated with microwave process such as saving of time, energy savings, and an enhancement in the metallurgical and mechanical properties in comparison with parts/components processed with the conventional casting techniques. However, microwaves were employed mostly for metal powder sintering in its recent years. Investigations into metallic-based material sintering yield good outcomes, which have motivated researchers to design, develop, and analyze possibilities of new techniques. Consequently, MC process come into existence and is gaining popularity. This means that MC is a growing field of research and development with gaps required to be filled to make it a future-ready process. But better understanding and knowledge of the microwave physics and the microwave process parameters will enlighten the process knowledge, which leads to better process control with quality castings. To make MC commercialization in the future, there is a need of further improvements in the process, which will attract the public/private-sector industries to invest more. In the future, the MC of ferrous materials will be explored also. A bright future awaits.

References

[1] Awida, M.H., Shah, N., Warren, B., Ripley, E., and Fathy, A.E., 2008, June. Modeling of an industrial microwave furnace for metal casting applications. In *2008 IEEE MTT-S International Microwave Symposium Digest* (pp. 221–224). IEEE, Elkridge, MD.

[2] Vasudev, H., Singh, G., Bansal, A., Vardhan, S., and Thakur, L., 2019. Microwave heating and its applications in surface engineering: a review. *Materials Research Express*, 6(10), p. 102001.

[3] Mishra, R.R., and Sharma, A.K., 2016. A review of research trends in microwave processing of metal-based materials and opportunities in microwave metal casting. *Critical Reviews in Solid State and Materials Sciences*, 41(3), pp. 217–255.

[4] Samyal, R., Bagha, A.K., and Bedi, R., 2020. The casting of materials using microwave energy: A review. *Materials Today: Proceedings*, 26, pp. 1279–1283.

[5] Aravindan, S., and Krishnamurthy, R., 1999. Joining of ceramic composites by microwave heating. *Materials Letters*, 38(4), pp. 245–249.

[6] Gupta, M., and Wong, W.L.E., 2005. Enhancing overall mechanical performance of metallic materials using two-directional microwave assisted rapid sintering. *ScriptaMaterialia*, 52(6), pp. 479–483.

[7] Mishra, R.R., and Sharma, A.K., 2019. Microstructural characteristics and tensile properties of in-situ and ex-situ microwave casts of Al-7039 alloy. *Materials Research Express*, 6(12), p. 126591.

[8] Nishitani, T., and Nippon Steel Corp, 1979. Method for sintering refractories and an apparatus therefor. U.S. Patent 4,147,911.

[9] Saitou, K., 2006. Microwave sintering of iron, cobalt, nickel, copper and stainless steel powders. *ScriptaMaterialia*, 54(5), pp. 875–879.

[10] Anklekar, R.M., Bauer, K., Agrawal, D.K., and Roy, R., 2005. Improved mechanical properties and microstructural development of microwave sintered copper and nickel steel PM parts. *Powder Metallurgy*, 48(1), pp. 39–46.

[11] Kubel, E., 2005. Advancement in microwave heating technology. *Ind. Heat.*, 62, p. 43.

[12] Gedevanishvili, S., Agrawal, D.K., Roy, R., Vaidhyanathan, B., and Penn State Research Foundation, 2003. Microwave processing using highly microwave absorbing powdered material layers. U.S. Patent 6,512,216.

[13] Kim, J., Mun, S.C., Ko, H.U., Kim, K.B., Khondoker, M.A.H., and Zhai, L., 2012. Review of microwave assisted manufacturing technologies. *International Journal of Precision Engineering and Manufacturing, 13*(12), pp. 2263–2272.

[14] Das, S., Mukhopadhyay, A.K., Datta, S., and Basu, D., 2009. Prospects of microwave processing: An overview. *Bulletin of Materials Science, 32*(1), pp. 1–13.

[15] Oghbaei, M., and Mirzaee, O., 2010. Microwave versus conventional sintering: A review of fundamentals, advantages and applications. *Journal of Alloys and Compounds, 494*(1–2), pp. 175–189.

[16] Chandrasekaran, S., Ramanathan, S., and Basak, T., 2012. Microwave material processing: A review. *AIChE Journal, 58*(2), pp. 330–363.

[17] Rybakov, K.I., Olevsky, E.A., and Krikun, E.V., 2013. Microwave sintering: Fundamentals and modeling. *Journal of the American Ceramic Society, 96*(4), pp. 1003–1020.

[18] Mishra, R.R., Rajesha, S., and Sharma, A.K., 2014. Microwave sintering of pure metal powders: A review. *Int. J. Adv. Mech. Eng, 4*(3), pp. 2250–3234.

[19] Sharma, A.K., Srinath, M.S., and Kumar, P., 2009. Microwave joining of metallic materials, Indian patent, Application no. 1994/Del/2009.

[20] Sharma, A.K., and Gupta, D., 2010. A method of cladding/coating of metallic and non metallic powders on metallic substrates by microwave irradiation, Indian patent, Application no. 527/Del/2010.

[21] Moore, A.F., Donald, E.S., and Marvin, S.M., 2006. Method and apparatus for melting metals, US patent, 7011136.

[22] Warren, B., Awida, M.H., and Fathy, A.E., 2012. Microwave heating of metals. *IET Microwave. Antennas Propag. 6,* 196 (2012).

[23] Wiedenmann, O., Ramakrishnan, R., Kiliç, E., Saal, P., Siart, U., and Eibert, T.F., 2012, March. A multi-physics model for microwave heating in metal casting applications embedding a mode stirrer. In *2012 the 7th German Microwave Conference* (pp. 1–4). IEEE, Elkridge, MD.

[24] Pehrson, B.P., and Moore, A.F., 2014. Method and mold for casting thin metal objects, U.S. patent, 870803.

[25] Pehrson, B.P., and Moore, A.F., 2015. Method for casting thin metal object, U.S. patent, 9004148.

[26] Mishra, R.R., and Sharma, A.K., 2015. A new in-situ casting technique using microwave energy at 2.45 GHz. In *Proc. Of the India International Science Festival- Young Scientists' Meet, DST, Government of India*, Design 58, 2015, pp. 1–7.

[27] Lingappa, M.S., Srinath, M.S., and Amarendra, H.J., 2017. Microstructural and mechanical investigation of aluminium alloy (Al 1050) melted by microwave hybrid heating. *Materials Research Express, 4*(7), p. 076504.

[28] Mishra, R.R., and Sharma, A.K., 2016. On mechanism of in-situ microwave casting of aluminium alloy 7039 and cast microstructure. *Materials & Design, 112*, pp. 97–106.

[29] Mishra, R.R., and Sharma, A.K., 2017. Structure-property correlation in Al-Zn-Mg alloy cast developed through in-situ microwave casting. *Materials Science and Engineering: A, 688*, pp. 532–544.

[30] Mishra, R.R., and Sharma, A.K., 2018. Effect of solidification environment on microstructure and indentation hardness of Al-Zn-Mg alloy casts developed using microwave heating. *International Journal of Metalcasting, 12*(2), pp. 370–382.

[31] Mishra, R.R., and Sharma, A.K., 2017. Effect of susceptor and mold material on microstructure of in-situ microwave casts of Al-Zn-Mg alloy. *Materials & Design*, *131*, pp. 428–440.

[32] Chandrasekaran, S., Basak, T., and Ramanathan, S., 2011. Experimental and theoretical investigation on microwave melting of metals. *Journal of Materials Processing Technology*, *211*(3), pp. 482–487.

[33] Mishra, R.R., and Sharma, A.K., 2018, April. Experimental investigation on in-situ microwave casting of copper. In *IOP Conference Series: Materials Science and Engineering* (Vol. 346, No. 1, p. 012052). IOP Publishing, Bristol, UK.

[34] Marahadige, S.L., Sridharmurthy, S.M., Jayraj, A.H., Mahabaleshwar, U.S., Lorenzini, G., and Lorenzini, E., 2018. Development of copper alloy by microwave hybrid heating technique and its characterization. *Journal Homepage: Http://iieta.org/Journals/IJHT*, *36*(4), pp. 1343–1349.

[35] Singh, S., Gupta, D., and Jain, V., 2016. Novel microwave composite casting process: Theory, feasibility and characterization. *Materials & Design*, *111*, pp. 51–59.

[36] Singh, S., Gupta, D., and Jain, V., 2018. Microwave melting and processing of metal: Ceramic composite castings. *Proceedings of the Institution of Mechanical Engineers, Part B: Journal of Engineering Manufacture*, *232*(7), pp. 1235–1243.

[37] Singh, S., Gupta, D., and Jain, V., 2018. Processing of Ni-WC-8Co MMC casting through microwave melting. *Materials and Manufacturing Processes*, *33*(1), pp. 26–34.

[38] Mishra, R.R., and Sharma, A.K., 2017. On melting characteristics of bulk Al-7039 alloy during in-situ microwave casting. *Applied Thermal Engineering*, *111*, pp. 660–675.

[39] Singh, S., Gupta, D., and Jain, V., 2019. Microwave processing and characterization of nickel powder based metal matrix composite castings. *Materials Research Express*, *6*(8), p. 0865b1.

[40] Campbell, J., 2015. *Complete Casting Handbook: Metal Casting Processes, Metallurgy, Techniques and Design*. Butterworth-Heinemann, Oxford.

[41] Ravi, B., 2005. *Metal Casting: Computer-Aided Design and Analysis*. PHI Learning Pvt. Ltd., Delhi, India.

[42] Eskin, D.G., and Katgerman, L., 2004. Mechanical properties in the semi-solid state and hot tearing of aluminium alloys. *Progress in Materials Science*, *49*(5), pp. 629–711.

[43] Loharkar, P.K., Ingle, A., and Jhavar, S., 2019. Parametric review of microwave-based materials processing and its applications. *Journal of Materials Research and Technology*, *8*(3), pp. 3306–3326.

[44] Chakrabarti, D.J., and Laughlin, D.E., 2004. Phase relations and precipitation in Al-Mg-Si alloys with Cu additions. *Progress in Materials Science*, *49*(3–4), pp. 389–410.

[45] Agrawal, D., 2006, August. Microwave sintering, brazing and melting of metallic materials. In *Sohn International Symposium; Advanced Processing of Metals and Materials Volume 4: New, Improved and Existing Technologies: Non-Ferrous Materials Extraction and Processing* (Vol. 4, pp. 183–192). John Wiley & Sons, Hoboken, NJ.

[46] Xu, L., Srinivasakannan, C., Peng, J., Guo, S., and Xia, H., 2017. Study on characteristics of microwave melting of copper powder. *Journal of Alloys and Compounds*, *701*, pp. 236–243.

[47] H.-M. Hu. Development of research on ZA27 alloy-A review. *Mater. Rev.* 12, no. 17 (1998).

[48] J.D. Rutherford, (1985). ZA alloy die casting, ZA casting alloys conference. *Proc. of Materials Science Forum 5*, (1985), pp. 33–36.

[49] Li, R.X., Rong-de, L.I., and Bai, Y.H., 2010. Effect of specific pressure on microstructure and mechanical properties of squeeze casting ZA27 alloy. *Transactions of Nonferrous Metals Society of China*, 20(1), pp. 59–63.

[50] Chen, T.J., Hao, Y., Sun, J., and Li, Y.D., 2003. Effects of Mg and RE additions on the semi-solid microstructure of a zinc alloy ZA27. *Science and Technology of Advanced Materials*, 4(6), p. 495.

[51] Aashuri, H., 2005. Globular structure of ZA27 alloy by thermomechanical and semi-solid treatment. *Materials Science and Engineering: A, 391*(1–2), pp. 77–85.

[52] Ashouri, S., Nili-Ahmadabadi, M., Moradi, M., and Iranpour, M., 2008. Semisolid microstructure evolution during reheating of aluminum A356 alloy deformed severely by ECAP. *Journal of Alloys and Compounds, 466*(1–2), pp. 67–72.

[53] Chen, Q., Yuan, B., Zhao, G., Shu, D., Hu, C., Zhao, Z., and Zhao, Z., 2012. Microstructural evolution during reheating and tensile mechanical properties of thixoforged AZ91D-RE magnesium alloy prepared by squeeze casting: Solid extrusion. *Materials Science and Engineering: A, 537*, pp. 25–38.

Chapter 11

Processing Biomaterials Using Microwave Energy and Its Futuristic Scopes

Shivani Gupta, Apurbba Kumar Sharma, and Dinesh Agrawal

Contents

11.1 Introduction

In the present scenario, energy consumption increases day by day in the industrial, transportation, commercial, and residential sectors. The industrial sector consumes more than 30% of the total energy consumed worldwide [1]. Therefore, there is an immense demand for low-energy-consuming materials processing techniques for all kinds of products. Figure 11.1 shows the worldwide energy consumption in various sectors by 2019 [2]. In this regard, microwave-assisted materials processing techniques can play a significant role in the manufacturing industry. It can reduce energy consumption and processing time, improve the properties of processed materials, and hence can be cost-effective and, more important, clean and environmentally friendly. Originally, microwaves were used in communication during World War II followed by in the food, wood, paper, leather, and rubber industries [3]. Although the use of microwave energy to process a wide variety of

DOI: 10.1201/9781003248743-11

metals, ceramics, polymers, and composites offer many new and exciting opportunities to researchers [4]. In metals, polymers, and composites, the use of microwave energy is still under development. On the other hand, many refractory ceramics are well processed using this technique. Microwave heating depends on the electrical and magnetic properties of a material to be processed. What decides how microwaves would interact with the targeted material under which mechanism?

Microwaves are electromagnetic waves and can be coupled with various materials differently depending on their dielectric and magnetic losses. Hence, it is a material property. This salient feature of microwave processing makes it fundamentally different from other existing heating sources like conventional furnace, laser, plasma, and electron beam. In microwave heating, the material core gets heated first and then the surface, while in

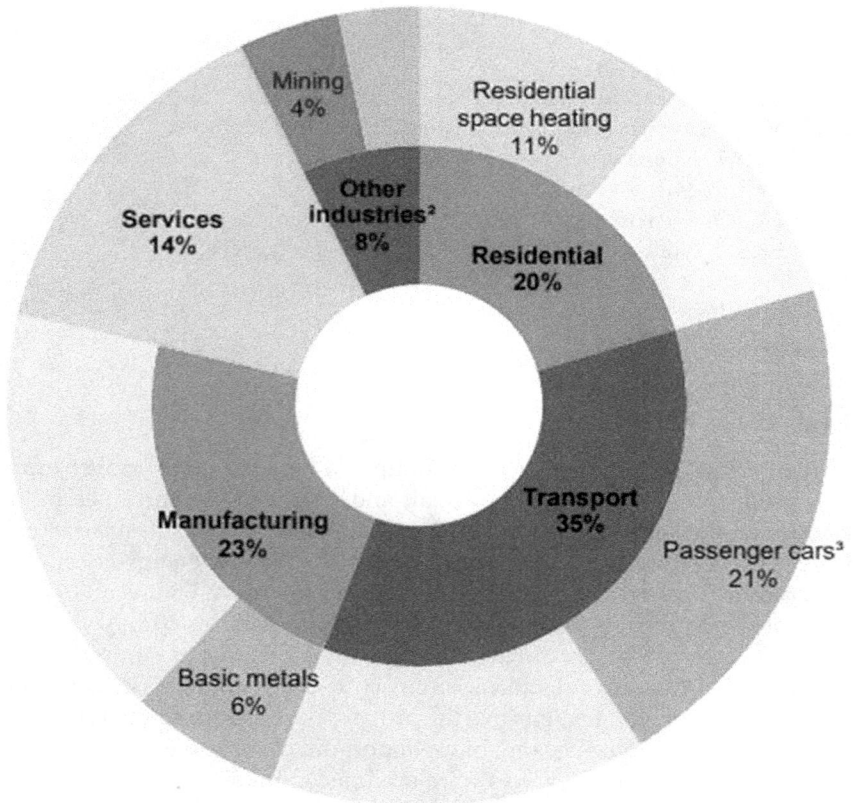

Figure 11.1 Worldwide energy consumption in various sectors by 2019.

other heating sources, heat is applied externally to the surface and then the interior surface gets heated through conduction or radiation. This heating mechanism reduces the processing time and energy consumption in microwave materials processing techniques by providing volumetric heating.

Microwave heating of metallic materials depends on the skin depth of the material, which is a function of permeability, permittivity, frequency of microwaves and electrical conductivity of a targeted material. Bulk metals having high conductivity that is associated with a short skin depth causes the reflection of most microwaves without absorption and transmission. However, powder metal can absorb microwaves very effectively due to the large surface area. In the case of ceramics, most of them are insulators with low conductivity and good dielectric loss that show significant microwave energy absorption and rapid heating. Hence, microwave-based manufacturing methods are widely used in processing of many specialty ceramics, which are challenging to manufacture conventionally.

Most polymeric materials and certain glass ceramics are thermal/electrical insulators with low dielectric loss and are mostly transparent to microwave. They allow the almost full transmission of the microwaves; hence, they only can be used as safety in microwave kitchen ovens. Many polymers were not processed much through microwaves as they are easy to process and require low heat for processing at the desired temperature. However, researchers have been exploring polymers and ceramics composites through microwave energy with the help of hybrid heating in which reinforcement and matrix behave differently and assist in heating the polymer through conduction and radiation [5, 6].

Microwave materials processing is extensively being used to process different composites involving metals, ceramics, and polymers as a matrix. The applications of such composites are automobiles, aerospace, power generation, and medical industries. Only materials with medical applications processed through microwaves are the most relevant for discussion in this chapter. Metal-, ceramic-, polymer- and composite-based biomaterials are being used for various medical devices and body implants as artificial biomaterials. In this chapter, we limit our discussion only to these materials.

11.2 Biomaterials and Their Classification

The use of certain materials as constituents of surgical implants is not new. Substitutions of human bone material for repairing seriously damaged body tissues have been reported since the pre-Christian era. Copper, bronze, iron, and gold were primarily utilized in the assembly of fractured bone. Later on, many medical practitioners and researchers developed new materials, which were more compatible with the human body and derived from natural or synthetic sources. However, the incidence of orthopedic interventions has increased enormously among the population since the beginning of the industrial era. The use of self-operating machines and the necessity for manual intervention are the leading causes of the ever-increasing number of accidents as production rhythm intensifies and interactions multiply. Furthermore, the moral awareness of everybody's right to existence has improved living standards and led to social and medical security improvements. These improvements draw materialists' attention to developing new biomaterials with better performance.

Biomaterial can be synthetic or natural sourced material suitable for constructing artificial organs and prostheses and replacing bone and tissues. It is a substance that has been engineered to interact with biological systems for a medical purpose, either a therapeutic or surgery. After inserting inside the human body, they interact with biological tissues and help in the growth of diseased tissues. They can be originated from natural sources and artificially developed from metals, ceramics, polymers, and composites. Figure 11.2 shows a classification of biomaterials and their resources. Artificial biomaterial applications are heart valves, orthopedic prostheses, hydrogel contact lenses, skin implants, plastic surgeries, and so forth.

Nowadays, researchers are very much interested in this field for being an interdisciplinary research domain with high-impact factors. A lot of work has been done on developing various artificial biomaterials and enhancing their quality and performance during service. However, there is a long way to find suitable biomaterial with the least adverse consequences. Therefore, a tremendous amount of research collaboration between research organizations and medical practitioners is being carried out. In the first place, surgeons have been more successful in reconstructing or replacing hard tissues like bone and teeth. Generally, metals, ceramics, polymers, and composites are commonly used in the reconstructive surgery of bones. Any material applicable for surgical implantation may cause a broad spectrum of

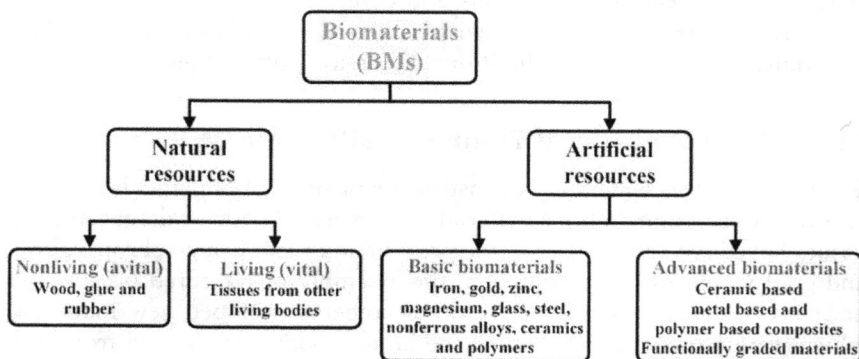

Figure 11.2 Primary classification of biomaterials and their resources.

biochemical reactions inside the body. This spectrum can be roughly divided into three categories:

1. Bioinert materials with minimal chemical reactivity
2. Complete biodegradable materials with possible dissolution into metabolic elements
3. Materials with controlled-reactive surfaces

Few metals and their alloys are (1) such as steel, titanium and titanium alloys, and cobalt-based alloys. They are used for medical devices and implants such as rods, pins, screws, sutures, and bone plates. In addition, ceramic, alumina, and zirconia are bioinert and used to replace hip and knee implants. The inertness of such ceramics, their high wear resistance, and their biocompatibility make them the best choice for dental implants. On the other hand, magnesium and its alloys, polymers (polylactic acid (PLA), polyglycolic acid (PGA), etc.), bioglass, and calcium apatites are material types (2). They are mainly used as artificial materials needed only to heal the diseased tissue and then degrade after service. Drug delivery systems and fixation aids are the leading applications of such materials. Materials coated with high wear-resistant materials to enhance the surface properties come under material type (3). These types of material are known as functionally graded material (FGM). Some polymers also fall under this category which is used in controlled reactive surface applications. The chronological development of biomaterials is shown in Figure 11.3.

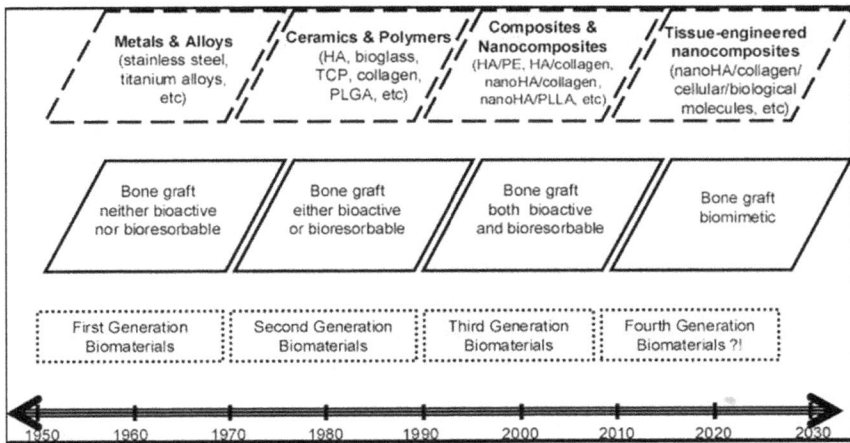

Figure 11.3 The chronological development of biomaterials from 1950 to 2030 as per three categories: bioinert, bioactive and controlled reactivity [42].

11.2.1 Metals and Alloys

Since ancient times, metals have been used as artificial biomaterials in grafting fractured body parts. Gold, copper, and bronze were the primary metals used in medical applications. However, chronological development in biomaterials involves many new metals such as stainless steel, Co-based alloys, Ti and Ti alloys, and Mg and Mg alloys. Earlier, stainless steel and Co-based alloys were widely used as biomaterials due to the presence of iron, cobalt, manganese, molybdenum, and sulfur as alloying elements [3]. All these elements are present in the human body as trace elements with specific biochemical reactions inside the body. However, Ni is present in most steels and leaches out after a long time of service and creates toxicity and inflammability responses in the body. It leads that the second surgical operation(s) usually becomes expected to prevent long-term exposure of the body to the toxic implant contents. Nowadays, titanium and titanium alloys are the most suitable candidates for dental, knee, and hip implants. The biocompatibility of titanium is more appropriate for the human body than steel and Co-based alloys. However, they exhibit unsatisfactory results, such as the stress shielding effect and metal ion releases. The stress shielding is caused due to significant differences in elastic modulus of implanted material and host tissues and, as a result, transfers different loads in implant and host tissue.

Therefore, these metallic materials are being reformed with biodegradable materials, including metal alloys, ceramic, polymers, and metallic glasses. As such, the nature of metallic biomaterials is transformed from bioinert to bioactive and multi-biofunctional (antibacterial, antiproliferation, anticancer, etc.). Foremost, magnesium is the most appropriate metal for biodegradable applications due to its many salient features. Magnesium is the essential mineral required in the human body for the proper functioning of other minerals that control the functioning of all organisms. In addition, magnesium has a high strength-to-weight ratio and elastic modulus equivalent to the human bone, eliminating stress shielding. Plus, magnesium has suitable mechanical properties and biocompatibility for bone implants.

11.2.2 Ceramics

Alumina, zirconia, bioglass, calcium phosphate ceramics and ceramic-polymer composites are preferably used for medical applications. Mainly, these bioceramics are used in hard-tissue transplantation. Alumina is used in dental implants due to its long life span and high wear resistance. Alumina is used for orthopedic prostheses, pacemakers and numerous prototypes of implantable devices. Zirconia is a white ceramic extensively used in hard-tissue restoration. White ceramic is the best substitute for teeth, commonly used in dentistry where inert biomaterial is desirable.

Bioglass and calcium phosphate ceramics are bioactive artificial materials that degrade after a definite service time. Orthopedic fixation aids, screws, bone plates and biodegradable rods are the applications of such ceramics. Incorporating drugs within and/or on the surface of the scaffold provides new chemical strategies that enhance the rate of healing of osteoporotic fractures, large bone defects, bone tumors, and bone infections. These requirements lead to biodegradable materials such as calcium phosphate ceramics, which are the most desirable material in drug delivery systems due to their unique and similar characteristics to human cortical bone. These are the calcium salts of phosphoric acid and contain oxide and hydroxide, forming calcium apatite. Calcium and phosphorous are essential minerals for the human body.

11.2.3 Polymers

Due to their unique properties, polymers are extensively used in medical applications, especially in soft-tissue implantation. A polymer is a substance that contains a molecular structure of many identical units bonded together; the single identical unit is called a monomer. They can be natural or synthetic; for example, plastic is the synthetic polymer used in daily life. And protein is a biological polymer that exists in the living body. Many polymers have been developed, but only such polymers have been used in medical applications and discussed in this chapter. There are many synthetic polymers primarily used in human body as implants. Polyetheretherketone (PEEK), PLA, PGA, polylactic-co-glycolic acid (PLGA), polycaprolactone (PCL), polyethylene (PE), polyethylene oxide (PEO), polyurethanes, polypropylene (PP), polystyrene (PS), polytetrafluoroethylene (PTFE), and polyethylene terephthalate (PET) are those polymers [38]. PEEK is the best polymer for hard-tissue transplants. Screws from PEEK are used for arthroplasty and dentistry. The acetabular cup of the hip joint from PEEK has high wear resistance and low weight. In short, PEEK has a long life span, tough material for orthopedic applications. PLA, PGA and PLGA are biodegradable polymers that are used in screws, rods, pins, and mesh/scaffold for implanting in the body and degrades completely within 6 to 12 months. PE is used for medical packing and catheter tubes due to the low coefficient of friction and PEO is used as biosensors. PP and PS are widely used to make surgical devices like syringes, surgical wares, test tubes, beakers, and others. PET is a synthetic fiber used to tie the cuts and surgical vessels. PTFE patch is mainly used for the treatment of abdominal hernias. Therefore, all these polymers have specific applications in the medical field.

11.2.4 Composites and FGMs

Composites are a combination of two or more materials and exhibit most likely better properties than their parent materials. Composites are of three

types: metal matrix, ceramic matrix, and polymer matrix in broad categories. In a metal matrix, the main constituent is metal and similarly, ceramic is
in ceramic matrix and polymer is in the polymer matrix. Many metal matrix
composites, especially Mg-based composites with metal and ceramic reinforcement such as Ti, Zn, TiO_2, MgO, ZnO, ZrO_2, TiB_2, and Al_2O_3 are widely
used in orthopedic implants. Likewise, ceramic-based composites of Al_2O_3,
ZrO_2, hydroxyapatite (HA), and bioglasses are used for hip, knee, and dental implants. Polymer-based composites are used in soft-tissue implantation.
Magnesium- and HA-based composites are biodegradable composites used
in drug delivery systems, scaffolds, screws, bone plates, and rods.

On the other hand, functionally graded materials are developed for a specific application with specific characteristics. These materials are fabricated
with the help of several materials processed under different processing conditions to achieve desired attributes. An acetabular cap of titanium alloys
coated with HA improves tissue growth at the hip implant. Coated metal
alloys, polymers, and composites are widely used as FGM with improved
wear resistance, corrosion resistance, and osseo-generation ability in the
actual working environment.

11.3 Microwave-Assisted Biomaterials Processing Methods

The application of microwaves in materials processing has demonstrated
many advantages in the fabrication of high-temperature materials. Microwave materials processing is a rapid heating technique in which a wide variety of materials are processed. Microwave heating has some very unique
characteristics: rapid internal and selective heating, fine microstructures,
and short heating cycles [6].

11.3.1 Microwave-Assisted Material Synthesis

Material synthesis can be liquid-phase and solid-phase synthesis methods in
which various precursor materials are used to synthesize the desired material
through multiple chemical reactions. Ceramics, metals, polymers, and composites are widely prepared using microwave-assisted synthesis processes for
medical applications. A considerable literature is available on the microwave-
assisted synthesis of HA, Mg-substituted HA, and a few polymers, as shown in
Table 11.1. Hydroxyapatite is extensively used in dental, orthopedic implants,
and drug delivery systems due to its high compatibility with human bone.

11.3.2 Microwave Heating and Sintering

Microwave heating is fundamentally different from conventional heating,
which involves radian/resistance and/or convection heating followed by transfer of thermal energy via conduction to the inside of the workpiece through a

Table 11.1 Published Data on the Microwave-Assisted Synthesis of Important Biomaterials

Precursor Materials	Processing Parameters (MW power and time) at 2.45 GHz	Synthesis Process	Particle Size/Shape	Remark	Reference
1. Calcium chloride 2. Sodium dihydrogen phosphate 3. Diammonium hydrogen phosphate	Domestic (1200 W) 5–10 min, pH=10	Ppt + filtered + additional MW (10 min)	–	The dipole present in a water molecule contributes the largest part of the energy required to overcome the energy barrier for the nuclei formation of HA crystal growth	[7]
1. Calcium nitrate tetrahydrate 2. Diammonium hydrogen phosphate	Domestic (780 W) 20 min, pH = 14	Ppt + drying in MW + Calcination (800 °C/3 h)+heat treatment at 1150 °C/3 h	–	Heat treatment (1150 °C) increased the crystallinity of HA by converting TCP and amorphous HA to crystalline HA	[8]
1. Gypsum 2. Diammonium hydrogen phosphate	MW (1200 W) 100 °C 5–120 min pH=8.2	Hydrothermal + washing + drying	Needle-like: L = 30–300 nm, D = 5–50 nm	A comparative study on conventional and MW heating in which the conversion of gypsum to HA reached 100% after 8 h (conventional heating) vs 5 min (MW heating)	[9]
1. Calcium nitrate tetrahydrate 2. β-glucose 3. Phosphoric acid ammonium hydroxide	Domestic (700 W) 5 min up to 2 h, pH = 10	Ppt + furnace heating up to 1200 °C	–	1. Crystals prepared in short time have more Ca disfigurement (less thermal stability). 2. Long MW exposure increases the degree of crystallinity of HA.	[10]
1. Calcium hydroxide derived from eggshells 2. Diammonium hydrogen phosphate	Domestic (800 W) Time and pH not mentioned	Ppt + washing repeatedly + dried overnight in an oven at 100 °C	Platelets: L = 33–50 nm, W = 8–14 nm	1. Achieved density = 3.12 g/cm³ and Ca/P ratio 1.67 2. Showed good sinterability	[11]

(Continued)

Table 11.1 (Continued)

Precursor Materials	Processing Parameters (MW power and time) at 2.45 GHz	Synthesis Process	Particle Size/Shape	Remark	Reference
1. Calcium nitrate tetrahydrate 2. Dibasic anhydrous sodium	Domestic (600 W) customized with a refluxing system 3 min, pH = 9	Suspension ppt+ Filtered + rinsed + dried (air at 200 °C for 4 h)	Rod-shaped and elliptical particles: ~10–25 nm	MW processing is an effective method, easy to repeat and can be optimized for mass production	[12]
1. Calcium nitrate tetrahydrate 2. Potassium dihydrogen phosphate	Domestic (1100 W) 40 min, pH = 9	Ppt (ultrasonic) + filtered + centrifugation (before MW) + milled (after MW)	Nanospheres = 35 nm	1. Mean particle size remained constant across all power settings of MW 20–100% 2. Mean particle size decreased when the ultrasound power increased during the synthesis process	[13]
1. Calcium nitrate 2. Disodium hydrogen phosphate 3. EDTA	Domestic (750 W) 33.3% power) 30 min, pH = 9–14	Ppt + MW + Centrifuged + washed + dried in vacuum + calcination	Rodlike: D = 200 nm	Nano-rods were synthesized and precursor was sintered above 700 °C	[14]
1. Calcium Nitrate 2. Potassium phosphate monobasic 3. Sodium nitrate, nitric acid and urea	Domestic (600 W) 5 min, pH = 1–3	MW-assisted solution combustion synthesis with Ca/P ratio = 1.5	Nanowhiskers: L = 1.2 mm D = 300–600 nm	1. An increased pH from the decomposition of urea served to drive the resultant CAP to be HA (stable at higher pH) 2. In this method, solution pH is essential to control the phase and morphology of CAP.	[15]
1. Eggshell- derived calcium hydroxide 2. Diammonium hydrogen phosphate	Domestic (900 W) 1. min pH = 10–11	Mixing + MW + centrifuge (3000 (rpm) + washed + dried	Agglomerated irregular round shape: D = 75 nm	The microwave-assisted method is economical, efficient, and repeatable	[16]

Reagents	Microwave condition	Process	Product	Remarks	Ref.
1. Calcium nitrate tetrahydrate 2. Ortho phosphoric acid	Domestic (800 W) 2 min pH=10	Ppt + MW + washed + calcination (500 °C)	Nanorods L = 37 nm D = 8 nm	Nano-HA and nanocomposite (HA/alumina) were prepared. HA cause less RBC lysis than its composite that could be used for orthopedic applications.	[17]
1. Calcium chloride 2. sodium carbonate and hydroxide 3. disodium hydrogen phosphate	Microwave applicator power not mentioned 3 min pH = 11	Mixing + MW + centrifuge + washed + dried	HA microspheres formed	The microwave-assisted method is economical, efficient and, repeatable and developed HA microspheres are suitable for drug delivery systems.	[18]
1. Calcium nitrate tetrahydrate 2. Orthophosphoric acid	Domestic (700 W) 1 min pH = 1–6	Ppt + dried + MW	Interwoven fibrous structure	Small size fibrous structure is found in microwave-assisted synthesis compared with conventional ones.	[19]
1. Diammonium hydrogen phosphate 2. Calcium nitrate tetrahydrate 3. Magnesium nitrate hexahydrate 4. Ammonium hydroxide	Domestic (600 W) 5 min pH=10	Ppt + MW + fiterd + washed + dried + sieved + heat-treated (900 °C l h)	Agglomerated irregular shaped particle	Mg substituted calcium phosphate bioceramic is prepared	[20, 21]
1. L-lactide 2. Chloroform, diethyl ether and methanol 3. Tetrahydrofuran 4. Dimethylformamide	Domestic (360 W) 3 min	Ppt + MW + filteration + dried	—	PLA (Mn = 7.6 × 104, yield >95%) is synthesized in only 20 min under microwave-assisted synthesis as compared to several hours in conventional synthesis.	[22]
1. Calcium nitrate tetrahydrate 2. Diammonium hydrogen phosphate	Commercial microwave (500 W) 20 min	Ppt + MW + filteration + washed + dried + sintered	Whiskers of HA with size 0.2 × 5 µm	Porous HA structure was prepared that can be used in drugs delivery system.	[23]

thermal conductivity mechanism. On the other hand, in microwave heating, the absorption of microwave energy is followed by volumetric heating involving the conversion of electromagnetic energy into thermal energy, and hence, it is very rapid and uniform. Thermal conduction in a conventional process is very sluggish, whereas in microwave heating, the role of thermal conductivity is minimized, and heating is instantaneous and is a function of the material's dielectric and magnetic properties. The heat is generated internally within the material instead of originating from external sources in a conventional furnace.

Microwave sintering is a well-known and well-developed sintering process in ceramic processing. It is a near near-net-shaped process that eliminates the postprocessing of the sintered products. At first, microwave sintering was used for refractory ceramics such as alumina and zirconia since it was difficult to sinter them using conventional sintering methods, long sintering periods, and high temperatures are employed. Alumina was the first bioinert ceramic sintered through microwave energy due to high sintering temperatures. Alumina of 98% density was processed in a very short processing time of 15 min at 1400°C through microwave sintering, and on the other hand, the same densification was achieved in conventional sintering at 1600°C in 2 h. Microwave-processed alumina required less energy consumption than conventional sintering and achieved better properties at lower sintering temperatures in a short time [24, 25]. Zirconia also is a bioceramic commonly used in making artificial teeth and is widely processed using microwave energy [26]. It needs a long processing time for full densification in the conventional process. Zirconia was sintered in a 2.45 GHz multimode microwave applicator at 1360°C in 2 min of processing time. Microwave sintered zirconia obtained a fine-grain structure of 0.25-μm size with 97.8% densification. In some more studies, Ca-stabilized zirconia was also sintered at 1400°C in 10 min using microwave and observed phase transformation from tetragonal to cubic phase. In conventional sintering, it takes 10 h to achieve the same degree of transformation and densification. The results also confirmed that microwave processing enhances the material properties by improving the kinetics of diffusion and phase changes [27–29]. Hydroxyapatite was sintered using microwave sintering and obtained good results [30].

In 1999, for the first time, Rustom Roy et al. reported sintering powder metals using microwave heating [31]. They successfully sintered stainless steel 316L and 434L, 316L is widely used in surgical instruments production and obtained better mechanical and microstructural properties than conventionally sintered samples [32]. Figure 11.4 shows typical microstructures of microwave sintered 316L and 434L stainless steels and sintering profiles of conventional and microwave sintering methods.

Ti alloys, especially Ti6Al4V, is the most suitable alloy for biomedical implants. Its composites were processed through the microwave-assisted sintering process [33]. Table 11.2 tabulates the results of microwave sintered

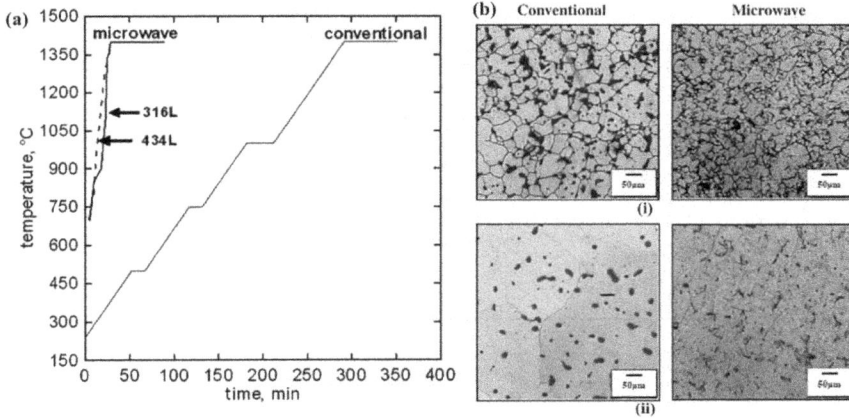

Figure 11.4 (a) Time–temperature curve of conventional and microwave sintering and (b) microstructural image of i-316L and ii-434L [32].

Table 11.2 Various Microwave-Sintered Biocomposites Used in Medical Implants.

Material	Processing	Outcomes	References
Mg/HAp	Microwave sintering	Improved YS, flexural, and compressive strength. HAp reduces hydrogen evolution and improves biocompatibility and bioactivity as compared to pure Mg	[34]
Mg/Y$_2$O$_3$	Microwave sintering followed by extrusion	Addition of yttria improved 0.2% YS modulus, UTS, ductility, and thermal stability	[35]
Mg/Bioglass	Microwave sintering	Improve mechanical properties along with reduced hydrogen evolution in vitro test, achieved better biocompatibility and cytotoxicity as compared to pure Mg	[36]
Al$_2$O$_3$/ZrO$_2$	Microwave, conventional sintering	Improved Relative density −99.35%, hardness −13.09 GPa, fracture toughness −11.62 MPam$^{1/2}$, bending strength 766.89 MPa as compared to conventional sintering	[37]
Ti/TiC porous composite	Microwave sintering	Sintered at 1620 °C for 2 min and achieved compressive strength 270.41 ± 24.97 MPa, YS 145.48±27.28 MPa, YM 10.97 ± 2.46 GPa, and hardness 545.4±13.9 HV higher than Ti alloy	[38]

(Continued)

Table 11.2 (Continued)

Material	Processing	Outcomes	References
Ti6Al4V/TiC/ HA	Microwave sintering	Sintered at 1620 °C for 2 min and achieved required porosity for bone implant. Developed composites showed better mechanical properties great biocompatibility, bioactivity, and osteoconductivity	[39, 40]
PP/MWCNT/ HA	Microwave sintering	Achieved higher modulus at higher HA%, transparent PP polymers with HA have good feasibility with MS and are useful for tissue engineering	[41]

various biocomposites studies. Magnesium and HA are bioactive metals and ceramics, mainly used in scaffolds, degradable screws, rods, and plates. A large amount of published literature exists on Mg/HA composites. However, this field still needs to explore using new cost-effective manufacturing techniques like microwave heating.

11.3.3 Microwave Coating

Coating is a postprocessing step to enhance the developed material's wear, corrosion, and surface properties. There are many well-established coating techniques: sol-gel, electrochemical deposition, electrophoretic deposition, plasma spraying, sputter coating, hot isostatic pressing, pulse laser deposition, and biomimetic synthesis [42]. But microwave coating is a new coating technique, which is in the development stage and is less explored. In 2002, alumina + titania ceramic composite was deposited on low carbon steel using a plasma spray method, and microwave irradiation was used as postprocessing for improving the wear resistance and corrosion resistance with fine microstructures [43]. Later, HA bioactive coating on Ti substrate was deposited by microarc oxidation and microwave hydrothermal posttreatment. The coated substrate was characterized for its physical properties, in vitro and in vivo tests, and observed that the HA layer enhances the bioactivity and bone contact interface and new bone formation. Similarly, many studies have been conducted to deposit apatite on metal implants, especially Ti and Ti6AL4V alloy and Mg alloys, to increase their bioactivity, and they show better hydrophilicity, physical properties, bioactivity, cell adhesion, and growth capability than uncoated ones [44–50].

11.3.4 Microwave-Assisted Additive Manufacturing

Microwave-assisted additive manufacturing is a novel and newly emerging method in biomaterials processing. In this method, materials are developed through an additive manufacturing process, also known as 3D printing, followed by microwave-assisted heating. M.A. Willert et al. explored the laser-assisted microwave plasma processing (LAMPP) technique to develop Al_2O_3 and ZrO_2 and analyzed the LAMPP process with laser sintering. As a result, LAMPP was proposed as a versatile processing technique for ceramic coating and sintering and melting metals and ceramics [51]. Sam Buls et al. processed Al_2O_3 reinforced with ZrO_2 through a microwave-assisted selective laser melting process. High-temperature ceramic was developed with high densities without any crack and thermal-runaway effect [52]. Hugo Curto et al. tried to prepare Al_2O_3 using additive manufacturing (stereolithography) process and microwave sintering. They achieved better relative density, elastic modulus, and microhardness than conventional heating and saved 30% processing time in microwave heating [53]. Very few reports are available in this domain. Therefore, there is ample scope to explore this area for the development of various materials including biomaterials.

11.4 Futuristic Scopes in Biomaterials Processing

Microwave-assisted biomaterials processing techniques are in their infancy. Many published reports are available in reputed journals on microwave-assisted processing of various biomaterials with improved material properties, confirming enhanced performance in the artificially developed working environmental conditions, such as an in vitro test. However, there are many challenges in processing biomaterials microwaves due to inherent limitations in microwave industrialization such as the nonuniform properties of the materials, variations in processing temperature, and the like. Biomaterials require a porosity gradient in drug delivery system that is difficult to produce with adequate strength. The simulation of the process is quite difficult due to the lack of data availability. These challenges are categorized as challenges in the microwave heating process, processing materials, tooling design, and controlling process parameters. Figure 11.5 depicts various associated challenges in microwave processing. These include nonuniform heating, hot spots, thermal runaway, and control of rapid heating, among others. However, various researchers are working to overcome such challenges in microwave heating.

Such challenges offer opportunities for the development and advancement of any field. The challenges shown in Figure 11.5 provide great scopes of research in microwave processing of biomaterials in the future. These challenges can be rectified by exploring advanced interdisciplinary research

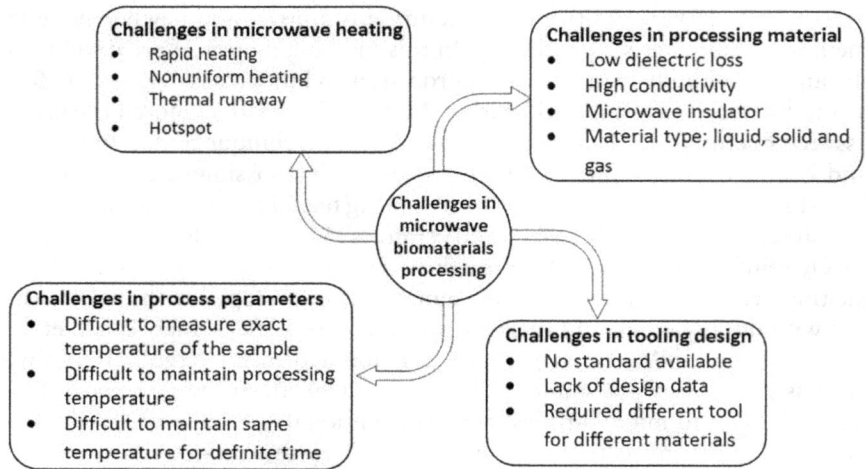

Figure 11.5 Challenges associated with microwave processing biomaterials.

like modeling and simulation of the process and tool design to be excellent scopes for the future, shown in Figure 11.6. Biocomposites can be developed using hybrid heating method by adopting appropriate tool design and furnace setup. Measurement of dielectric properties of the targeted biomaterials through vector network analysis (VNA) can be an excellent opportunity to design proper tools and setup according to their dielectric characteristics.

11.5 Conclusion

This chapter has been focused on biomaterials processing using the application of newly emerging field of microwave-assisted sintering method. Microwave materials processing is a well-established technique in processing ceramics with improved product quality. However, this process needs to be further explored in the development of certain biomaterials. Most biomaterials of interest are prepared by combining metals, ceramics, polymers, and their composites to obtain the desired characteristics suitable for biomedical applications. Stainless steel, Mg and Mg alloys, and Ti and Ti alloys are the most common ingredients of biomaterials. Similarly, alumina, zirconia, HA, and bioglasses are also used for orthopedic and dental applications. In addition, polymers and their composites are widely used for surgical devices and soft tissue implantation. Composites and FGM are developed for specific applications with specific properties.

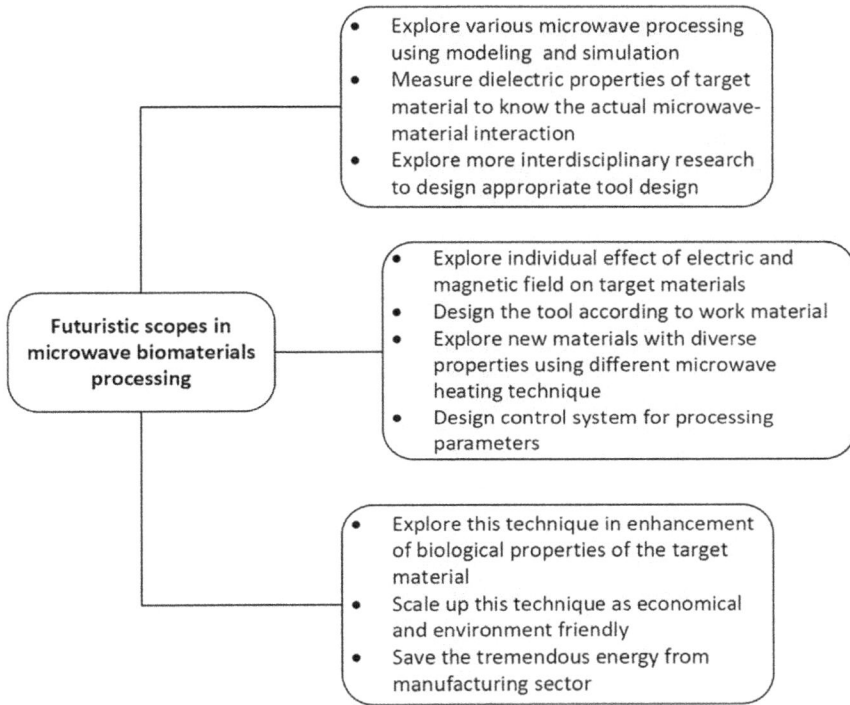

Figure 11.6 Futuristic scopes in microwave biomaterials processing.

In this chapter, only microwave-processed biomaterials have been discussed with the help of available literature. A large amount of work has been carried out on the microwave synthesis of biomaterials, and it still continues obtaining the desired properties. Likewise, microwave sintering is also a well-recognized technique in bioceramic and composites processing. On the other hand, microwave coating is a new technique used in the coating of apatite on metal alloy implants for enhancing their surface properties and wear and corrosion resistance. There are many future scopes to explore in different biomaterials along with conventional coating techniques followed by microwave sintering.

Finally, there are still many unresolved challenges associated with microwave processing biomaterials so that readers can find the opportunity for future work. These challenges only offer scopes of future work in this research field.

This book chapter is dedicated to all who would be reading this!

References

[1] Ahmad, Tanveer, and Dongdong Zhang. "A critical review of comparative global historical energy consumption and future demand: The story told so far." *Energy Reports* 6 (2020): 1973–1991.

[2] www.iea.org/data-and-statistics/data-product/energy-efficiency-indicators (visited on 10/11/2021).

[3] Sharma, Apurbba Kumar, and Shivani Gupta. "Microwave processing of biomaterials for orthopedic implants: Challenges and possibilities." *JOM* 72, no. 3 (2020): 1211–1228.

[4] Agarwal, D. K., J. Cheng, Y. Fang, and R. Roy. "Microwave processing of ceramics, composites and metallic materials." *Microwave Solutions for Ceramic Engineers*, 205–228. The American Ceramic Society: Ohio, 2005.

[5] Mishra, Radha Raman, and Apurbba Kumar Sharma. "Microwave: Material interaction phenomena: heating mechanisms, challenges and opportunities in material processing." *Composites Part A: Applied Science and Manufacturing* 81 (2016): 78–97.

[6] Gupta, Shivani, and Apurbba Kumar Sharma. "Sintering of biomaterials for arthroplasty: A comparative study of microwave and conventional sintering techniques." *Applied Mechanics and Materials* 895 (2019): 83–89. Trans Tech Publications Ltd.

[7] Horikoshi, Satoshi, Robert F. Schiffmann, Jun Fukushima, and Nick Serpone. *Microwave Chemical and Materials Processing*. Springer: Singapore, 2018.

[8] Lerner, E., and S. Sarig. "Enhanced maturation of hydroxyapatite from aqueous solutions using microwave irradiation." *J. Mater. Sci. Mater. Med.* 2 (1991): 138–141.

[9] Kundu, P. K., T. S. Waghode, D. Bahadur, and D. Datta. "Cell culture approach to biocompatibility evaluation of unconventionally prepared hydroxyapatite." *Med. Biol. Eng. Comput.* 36 (1998): 654–658.

[10] Katsuki, H., S. Furuta, and S. Komarneni. "Microwave- versus conventional-hydrothermal synthesis of hydroxyapatite crystals from gypsum." *J. Am. Ceram. Soc.* 82 (1999): 2257–2259.

[11] Yang, Z., Y. Jiang, Y. Wang, L. Ma, and F. Li. "Preparation and thermal stability analysis of hydroxyapatite derived from the precipitation process and microwave irradiation method." *Mater. Lett.* 58 (2004): 3586–3590.

[12] Siva Rama Krishna, D., A. Siddharthan, S. K. Seshadri, and T. S. Sampath Kumar. "A novel route for synthesis of nano-crystalline hydroxyapatite from eggshell waste." *J. Mater. Sci. Mater. Med.* 18 (2007): 1735–1743.

[13] Kalita, S. J., and S. Verma. "Nano-crystalline hydroxyapatite bioceramic using microwave radiation: Synthesis and characterization." *Mater. Sci. Eng. C.* 30 (2010): 295–303.

[14] Poinern, G., R. Brundavanam, X. T. Le, S. Djordjevic, M. Prokic, and D. Fawcett. "Thermal and ultrasonic influence in the formation of nano-meter scale hydroxyapatite bio-ceramic." *Int. J. Nano-Medicine* 6 (2011): 2083–2095.

[15] Mishra, V. K., S. K. Srivastava, B. P. Asthana, and D. Kumar. "Structural and spectroscopic studies of hydroxyapatite nano-rods formed via microwave-assisted synthesis route." *J. Am. Ceram. Soc.* 95 (2012): 2709–2715.

[16] Wagner, D. E., K. M. Eisenmann, A. L. Nestor-kalinoski, and S. B. Bhaduri. "A microwave-assisted solution combustion synthesis to produce europium doped calcium phosphate nano-whiskers for bioimaging applications." *ActaBiomater.* 9 (2013): 8422–8432.

[17] Sajahan, N. A., W. Mohd, and A. Wan. "Microwave irradiation of Nano-hydroxyapatite from chicken eggshells and duck eggshells." *Sci. World J.* 2014 (2014).

[18] Radha, G., S. Balakumar, B. Venkatesan, and E. Vellaichamy. Evaluation of hemocompatibility and in vitro immersion on microwave-assisted hydroxyapatite alumina nano-composites." *Mater. Sci. Eng. C.* 50 (2015): 143–150.

[19] Xiao, Wenqian, Haiming Gao, Moyuan Qu, Xue Liu, Jing Zhang, Hong Li, Xiaoling Yang, Bo Li, and Xiaoling Liao. "Rapid microwave synthesis of hydroxyapatite phosphate microspheres with hierarchical porous structure." *Ceramics International* 44, no. 6 (2018): 6144–6151.

[20] Sabu, Ummen, G. Logesh, Mohammad Rashad, Anand Joy, and M. Balasubramanian. "Microwave assisted synthesis of biomorphic hydroxyapatite." *Ceramics International* 45, no. 6 (2019): 6718–6722.

[21] Khan, Nida Iqbal, Kashif Ijaz, Muniza Zahid, Abdul S. Khan, Mohammed Rafiq Abdul Kadir, Rafaqat Hussain, Jawwad A. Darr, and Aqif A. Chaudhry. "Microwave assisted synthesis and characterization of magnesium substituted calcium phosphate bioceramics." *Materials Science and Engineering: C* 56 (2015): 286–293.

[22] Zhou, Huan, Timothy J. F. Luchini, and Sarit B. Bhaduri. "Microwave assisted synthesis of amorphous magnesium phosphate nanospheres." *Journal of Materials Science: Materials in Medicine* 23, no. 12 (2012): 2831–2837.

[23] Agrawal, Dinesh K., Y. Fang, Della M. Roy, and R. Roy. "Fabrication of hydroxyapatite ceramics by microwave processing." *MRS Online Proceedings Library (OPL)* 269 (1992).

[24] Singla, Pankil, Rajeev Mehta, Dusan Berek, and S. N. Upadhyay. "Microwave assisted synthesis of poly (lactic acid) and its characterization using size exclusion chromatography." *Journal of Macromolecular Science, Part A* 49, no. 11 (2012): 963–970.

[25] Agrawal, Dakshi, Vahid Tarokh, Ayman Naguib, and Nambi Seshadri. "Space-time coded OFDM for high data-rate wireless communication over wideband channels." *VTC'98. 48th IEEE Vehicular Technology Conference. Pathway to Global Wireless Revolution (Cat. No. 98CH36151)*, vol. 3, pp. 2232–2236. IEEE, New York, 1998.

[26] Brosnan, Kristen H., Gary L. Messing, and Dinesh K. Agrawal. "Microwave sintering of alumina at 2.45 GHz." *Journal of the American Ceramic Society* 86, no. 8 (2003): 1307–1312.

[27] Borrell, Amparo, María D. Salvador, Felipe L. Peñaranda-Foix, and Jose M. Cátala-Civera. "Microwave sintering of zirconia materials: Mechanical and microstructural properties." *International Journal of Applied Ceramic Technology* 10, no. 2 (2013): 313–320.

[28] Shukla, Mayur, Sumana Ghosh, Nandadulal Dandapat, Ashis K. Mandal, and Vamsi K. Balla. "Comparative study on conventional sintering with microwave sintering and vacuum sintering of Y 2 O 3-Al 2 O 3-ZrO 2 ceramics." *Journal of Materials Science and Chemical Engineering* 4, no. 2 (2016): 71.

[29] Binner, Jon, Ketharam Annapoorani, Anish Paul, Isabel Santacruz, and Bala Vaidhyanathan. "Dense nanostructured zirconia by two stage conventional/hybrid microwave sintering." *Journal of the European Ceramic Society* 28, no. 5 (2008): 973–977.

[30] Fang, Yi, Dinesh K. Agrawal, Della M. Roy, and Rustum Roy. "Microwave sintering of hydroxyapatite ceramics." *Journal of Materials Research* 9, no. 1 (1994): 180–187.

[31] Roy, Rustum, Dinesh Agrawal, Jiping Cheng, and Shalva Gedevanishvili. "Full sintering of powdered-metal bodies in a microwave field." *Nature* 399, no. 6737 (1999): 668–670.

[32] Panda, S. S., V. Singh, Agrawal Upadhyaya, and D. Agrawal. "Sintering response of austenitic (316L) and ferritic (434L) stainless steel consolidated in conventional and microwave furnaces." *ScriptaMaterialia* 54, no. 12 (2006): 2179–2183.

[33] Luo, S. D., C. L. Guan, Y. F. Yang, G. B. Schaffer, and M. Qian. "Microwave heating, isothermal sintering, and mechanical properties of powder metallurgy titanium and titanium alloys." *Metallurgical and Materials Transactions A* 44, no. 4 (2013): 1842–1851.

[34] Xiong, Guangyao, Yanjiao Nie, Dehui Ji, Jing Li, Chunzhi Li, Wei Li, Yong Zhu, Honglin Luo, and Yizao Wan. "Characterization of biomedical hydroxyapatite/magnesium composites prepared by powder metallurgy assisted with microwave sintering." *Current Applied Physics* 16, no. 8 (2016): 830–836.

[35] Tun, Khin Sandar, and M. Gupta. "Improving mechanical properties of magnesium using nano-yttria reinforcement and microwave assisted powder metallurgy method." *Composites Science and Technology* 67, no. 13 (2007): 2657–2664.

[36] Wan, Yizao, Teng Cui, Wei Li, Chunzhi Li, Jian Xiao, Yong Zhu, Dehui Ji, Guangyao Xiong, and Honglin Luo. "Mechanical and biological properties of bioglass/magnesium composites prepared via microwave sintering route." *Materials & Design* 99 (2016): 521–527.

[37] Zhou, Kui, Chunfa Dong, Xianglin Zhang, Lei Shi, Zhichao Chen, Yanlin Xu, and Hao Cai. "Preparation and characterization of nanosilver-doped porous hydroxyapatite scaffolds." *Ceramics International* 41, no. 1 (2015): 1671–1676.

[38] Veljovic, Dj, Ilmars Zalite, Eriks Palcevskis, I. Smiciklas, Rada Petrovic, and Dj Janackovic. "Microwave sintering of fine grained HAP and HAP/TCP bioceramics." *Ceramics International* 36, no. 2 (2010): 595–603.

[39] Choy, Man Tik, Chak Yin Tang, Ling Chen, Chi Tak Wong, and Chi Pong Tsui. "In vitro and in vivo performance of bioactive Ti6Al4V/TiC/HA implants fabricated by a rapid microwave sintering technique." *Materials Science and Engineering: C* 42 (2014): 746–756.

[40] Choy, Man-Tik, Chak-Yin Tang, Ling Chen, Wing-Cheung Law, Chi-Pong Tsui, and William Weijia Lu. "Microwave assisted-in situ synthesis of porous titanium/calcium phosphate composites and their in vitro apatite-forming capability." *Composites Part B: Engineering* 83 (2015): 50–57.

[41] Chen, Ling, Chak Yin Tang, Harry Siu-lung Ku, Chi Pong Tsui, and Xu Chen. "Microwave sintering and characterization of polypropylene/multi-walled carbon nanotube/hydroxyapatite composites." *Composites Part B: Engineering* 56 (2014): 504–511.

[42] Maitz, Manfred F. "Applications of synthetic polymers in clinical medicine." *Biosurface and Biotribology* 1, no. 3 (2015): 161–176.

[43] Asri, R. I. M., W. S. W. Harun, M. A. Hassan, S. A. C. Ghani, and Z. Buyong. "A review of hydroxyapatite-based coating techniques: Sol-gel and electrochemical depositions on biocompatible metals." *Journal of the Mechanical Behavior of Biomedical Materials* 57 (2016): 95–108.

[44] Sharma, Apurbba Kr, and R. Krishnamurthy. "Microwave processing of sprayed alumina composite for enhanced performance." *Journal of the European Ceramic Society* 22, no. 16 (2002): 2849–2860.

[45] Zhou, Huan, Maryam Nabiyouni, and Sarit B. Bhaduri. "Microwave assisted apatite coating deposition on Ti6Al4V implants." *Materials Science and Engineering: C* 33, no. 7 (2013): 4435–4443.

[46] Sadiq, Taoheed Olohunde, Nursyaza Siti, and Jamaliah Idris. "A study of strontium-doped calcium phosphate coated On Ti6Al4V using microwave energy." *Journal of Bio-and Tribo-Corrosion* 4, no. 3 (2018): 1–8.

[47] Yu, Nian, Shu Cai, Fengwu Wang, Feiyang Zhang, Rui Ling, Yue Li, Yangyang Jiang, and Guohua Xu. "Microwave assisted deposition of strontium doped hydroxyapatite coating on AZ31 magnesium alloy with enhanced mineralization ability and corrosion resistance." *Ceramics International* 43, no. 2 (2017): 2495–2503.

[48] Zhou, Huan, Shiqin Kong, Yan Pan, Zhiguo Zhang, and Linhong Deng. "Microwave-assisted fabrication of strontium doped apatite coating on Ti6Al4V." *Materials Science and Engineering: C* 56 (2015): 174–180.

[49] Du, Qing, Daqing Wei, Shaodong Wang, Su Cheng, Yaming Wang, Baoqiang Li, Dechang Jia, and Yu Zhou. "TEM analysis and in vitro and in vivo biological performance of the hydroxyapatite crystals rapidly formed on the modified microarc oxidation coating using microwave hydrothermal technique." *Chemical Engineering Journal* 373 (2019): 1091–1110.

[50] Jiang, Wenping, Jiping Cheng, and K. Dinesh. "Improved mechanical properties of nanocrystalline hydroxyapatite coating for dental and orthopedic implants." *Mater Res Soc* 1140 (2009): 1140-HH03.

[51] Willert-Porada, M. A., A. Rosin, P. Pontiller, C. Richter, and J. Boeckler. "Additive manufacturing of ceramic composites by laser assisted microwave plasma processing, LAMPP." In *2015 IEEE MTT-S International Microwave Symposium*, pp. 1–4. IEEE, New York, 2015.

[52] Buls, Sam, Jozef Vleugels, and Brecht Van Hooreweder. "Microwave assisted selective laser melting of technical ceramics." In *Proceedings of the 29th Annual International Solid Freeform Fabrication Symposium 2018-An Additive Manufacturing (AM) Conference*, pp. 2349–2357. University of Texas, Austin, TX, USA, 2018.

[53] Curto, Hugo, Anthony Thuault, Florian Jean, MaxenceViolier, Vedi Dupont, Jean-Christophe Hornez, and Anne Leriche. "Coupling additive manufacturing and microwave sintering: A fast processing route of alumina ceramics." *Journal of the European Ceramic Society* 40, no. 7 (2020): 2548–2554.

For Product Safety Concerns and Information please contact our EU
representative GPSR@taylorandfrancis.com
Taylor & Francis Verlag GmbH, Kaufingerstraße 24, 80331 München, Germany

www.ingramcontent.com/pod-product-compliance
Lightning Source LLC
Chambersburg PA
CBHW060552220326
41598CB00024B/3083